Professor Stewart's Casebook of Mathematical Mysteries

By the Same Author

Professor Stewart's Casebook of Mathematical Mysteries

Ian Stewart

PROFILE BOOKS

This paperback edition published in 2015
First published in 2014
PROFILE BOOKS LTD
3 Holford Yard
Bevin Way
London WC1X 9HD
www.profilebooks.com

10 9 8 7 6 5 4 3 2 1

Text design by Sue Lamble
Typeset in Stone Serif by Data Standards Ltd, Frome, Somerset.
Printed and bound in Britain by CPI Group (UK) Ltd, Croydon, CR0 4YY

A CIP catalogue record for this book is available from the British Library.

ISBN 978 1 84668 348 0
eISBN 978 1 84765 432 8

CONTENTS

• •

Acknowledgements

page 13 Left and centre figures: Laurent Bartholdi and André
Henriques. Orange peels and Fresnel integrals, *Mathematical
Intelligencer* 34 No. 4 (2012) 1–3.

page 13 Right figure: Luc Devroye.

page 27 Box puzzle concept: Moloy De.

page 40 Extract from *Not a Wake*: Mike Keith.

page 62 Haiku: Daniel Mathews, Jonathan Alperin, Jonathan
Rosenberg.

page 67 Figure: http://getyournotes.blogspot.co.uk/2011/08/
why-do-some-birds-fly-in-v-formations.html

page 71 Amazing squares: devised by Moloy De and Nirmalya
Chattopadhyay.

page 72 The Thirty-Seven Mystery: based on observations by
Stephen Gledhill.

page 74 Clueless pseudoku: Gerard Butters, Frederick Henle,
James Henle and Colleen McGaughey. Creating clueless
puzzles, *The Mathematical Intelligencer* 33 No. 3 (Fall 2011)
102–105.

page 96 Figure: Eric W. Weisstein, 'Brocard's Conjecture,' from
MathWorld—A Wolfram Web Resource: http://mathworld.
wolfram.com/BrocardsConjecture.html

page 101 Right Figure: Steven Snape.

page 109 Figure: Courtesy of the UW-Madison Archives.

page 123 Left figure: [George Steinmetz, courtesy of Anastasia
Photo].

page 123 Right figure: NASA, HiRISE on Mars Reconnaissance Orbiter.

page 124 Right figure: Rudi Podgornik.

page 125 Figure: Veit Schwämmle and Hans J. Herrmann. Solitary wave behaviour of sand dunes, *Nature* 426 (2003) 619–620.

page 131 Figure: Persi Diaconis, Susan Holmes and Richard Montgomery, Dynamical bias in the coin toss, *SIAM Review* 49 (2007) 211–223.

page 205 Figures: Joshua Socolar and Joan Taylor. An aperiodic hexagonal tile, *Journal of Combinatorial Theory Series A* 118 (2011) 2207–2231; http://link.springer.com/article/10.1007%2Fs00283-011-9255-y

pages 235–7 Figures: Michael Elgersma and Stan Wagon, Closing a Platonic gap, *The Mathematical Intelligencer* (2014) to appear.

The following figures are reproduced under the Creative Commons Attribution 3.0 Unported license and credited as requested on the description page:

page 95 Krishnavedala.

page 101 (left) Ricardo Liberato.

page 102 Tekisch.

page 139 Andreas Trepte, www.photo-natur.de.

page 180 Braindrain0000.

page 183 LutzL.

page 296 Walters Art Museum, Baltimore.

Introducing Soames and Watsup

∙ ∙

Professor Stewart's Cabinet of Mathematical Curiosities appeared in 2008, just before Christmas. Readers seemed to like its random mixture of quirky mathematical tricks, games, weird biographies, snippets of strange information, solved and unsolved problems, odd factoids, and the occasional longer and more serious piece on topics such as fractals, topology, and Fermat's Last Theorem. So in 2009 it was followed by *Professor Stewart's Hoard of Mathematical Treasures*, which continued in the same vein with an intermittent pirate theme.

They say that three is a good number for a trilogy. The late Douglas Adams of *The Hitchhiker's Guide to the Galaxy* fame did eventually decide that four was better and five better still, but three sounds like a good place to start. So, after a gap of five years, here is *Professor Stewart's Casebook of Mathematical Mysteries*. This time, however, there's a new twist. The short quirky items, such as Hexakosioihexekontahexaphobia, the Thrackle Conjecture, What Shape is an Orange Peel?, the RATS Sequence, and Euclid's Doodle, are still there. So are more substantial articles about solved and unsolved problems: Pancake Numbers, the Goldbach Conjecture, the Erdős Discrepancy Problem, the Square Peg Conjecture, and the ABC Conjecture. So are the jokes, poems, and anecdotes. Not to mention unusual applications of mathematics to flying geese, clumps of mussels, spotty leopards, and bubbles in Guinness. But these miscellanea are now interspersed with a series of narrative episodes featuring a Victorian detective and his medical sidekick—

I know what you're thinking. However, I developed the idea a year or so *before* Benedict Cumberbatch and Martin Freeman's spectacularly successful modern take on Sir Arthur Conan Doyle's much-loved characters hit the small screen. (Trust me.) More to the point, it's *not that pair*. Not even as portrayed in Sir Arthur's original stories. Yes, my guys live in the original time period, but *across the road* at number 222B. From there, they cast envious eyes on the stream of rich clients entering the premises of the more famous duo. And from time to time a case comes up that their illustrious neighbours have shunned or failed to solve: such arcane mysteries as the Sign of One, the Dogs that Fight in the Park, the Catflap of Fear, and the Greek Integrator. Then Hemlock Soames and Dr John Watsup put their brains in gear, show their true colours and their strength of character, and triumph over adversity and lack of market presence.

These are *mathematical* mysteries, you appreciate. Their solutions demand an interest in mathematics and an ability to think clearly, attributes in which Soames and Watsup are by no means deficient. These passages are signalled by the symbol $\mathcal{Q}$. Along the way we learn of Watsup's prior military career in Al-Jebraistan and Soames's battles with his arch-enemy Professor Mogiarty, inevitably leading to the final fatal confrontation atop the Schtickelbach Falls. And then—

It is fortunate that Dr Watsup recorded so many of their joint investigations in his memoirs and unpublished notes. I am grateful to his descendants Underwood and Verity Watsup for permitting me unprecedented access to family documents, and for generously granting me permission to include extracts here.

Coventry, March 2014

Note on Units

• •

In Soames and Watsup's era, the standard units of measurement in Britain were imperial, not metric as they mostly are today, and the currency was not decimal. American readers will have no problems with imperial units; admittedly, the gallon is different on either side of the Atlantic, but that unit of measurement doesn't appear anyway. To avoid inconsistencies I've used units appropriate to the Victorian era, even for topics that are not part of the Soames/Watsup canon, except when narrative imperative demands metric.

Here's a quick guide to the relevant units with metric/decimal equivalents.

Most of the time the actual unit doesn't matter: you could leave the numbers unchanged, but cross out 'inch' or 'yard' and replace it by an unspecified 'unit'. Or choose whichever seems convenient (metre for yard, for example).

Lengths

1 foot (ft) = 12 inches (in)	304·8 mm
1 yard (yd) = 3 feet	0·9144 m
1 mile (mi) = 1760 yards = 5280 feet	1·609 km
1 league (lea) = 3 miles	4·827 km

Weights

1 pound (lb) = 16 ounces (oz)	453·6 g

1 stone (st) = 14 pounds		6·35 kg
1 hundredweight (cwt) = 8 stone = 112 pounds		50·8 kg
1 ton (t) = 20 hundredweights = 2240 pounds		1·016 tonnes

Money

1 shilling (s) = 12 pence (d) [singular: penny]		5 new pence
1 pound (£) = 20 shillings = 240 pence		
1 sovereign = 1 pound (coin)		
1 guinea = £1.1s.		£1.05
1 crown = 5s.		25 new pence

Thruppeny bit = colloquial term for a three pence coin.

The Scandal of the Stolen Sovereign

The private detective took his wallet from his pocket, ascertained that it was still empty, and sighed. Standing at the window of his lodgings at number 222B he stared morosely across the street. The strains of an Irish air, expertly played on a Stradivarius, were just discernible above the clatter of passing carriages. Really, the man was *insufferable*! Soames stared at the stream of people entering the portals of his famous competitor. Most were wealthy members of the upper classes. Those that appeared not to be wealthy members of the upper classes were, with few exceptions, *representing* wealthy members of the upper classes.

Criminals just weren't committing the kind of crime that affected the sort of people who would engage the services of Hemlock Soames.

For the past two weeks, Soames had watched with envious eyes as client after client was ushered into the presence of the person they believed to be the greatest detective in the world. Or, at least, in London, which for Victorian England amounted to the same thing. Meanwhile his own doorbell remained mute, the bills piled up, and Mrs Soapsuds was threatening eviction.

He had only one case on the books. Lord Humphshaw-Smattering, owner of the Glitz Hotel, believed that one of his waiters had stolen a gold sovereign: value, one pound sterling. To be fair, Soames could do with a sovereign himself right now. But it was hardly the stuff to attract the sensationalist yellow press, upon whom, deplorable as it might be, his future depended.

Soames studied his case notes. Three friends, Armstrong, Bennett, and Cunningham, had partaken of dinner at the hotel, and had been presented with a bill for £30. Each had given the waiter Manuel ten gold sovereigns. But then the maître d' noticed that there had been an error, and the bill was actually £25. He gave the waiter five sovereigns to return to the men. Since £5 wasn't divisible by 3, Manuel suggested that he might keep two of the coins as a tip and give them back one sovereign each, hinting that they were fortunate to have any of the overpayment returned.

The customers agreed, and all was well until the maître d' noticed an arithmetical discrepancy. Now the men had each paid £9, a total of £27. Manuel had a further £2, making a total of £29.

One pound was missing.

Humphshaw-Smattering was convinced that Manuel had stolen it. Although the evidence was circumstantial, Soames knew that the waiter's livelihood depended on resolving the mystery. If Manuel were to be dismissed with a bad reference, he would never get another job.

Where did the missing pound go?

See page 249 for the answer.

. . . .

Number Curiosity*

In detective work, it is vital to be able to spot a pattern. Soames's unpublished and untitled monograph containing two thousand and forty-one instructive examples of patterns includes the following. Work out

11×91
11×9091
11×909091
11×90909091
11×9090909091

Soames would have used pen and paper, and modern readers

---

* Many items in this compilation that do not refer directly to criminal cases are extracted from handwritten notes, some of whose contents have been collected and published, with Soames's permission, as *Doctor Watsup's Vault of Forensic Anomalies*, and will be reproduced without further notification. Some are of later date, added by Watsup's literary executors, and the assiduous reader will instantly identify such anachronisms.

may do likewise if they can remember how. A calculator is always an option, but they tend to run out of digits. The pattern continues indefinitely: this can't be proved using a calculator, but it can be deduced from the old-fashioned method. So, *without* doing any further calculations, what is

$11 \times 9090909090909091$?

A harder question is: why does it work?

See page 250 for the answers.

Track Position

Lionel Penrose invented a variation on traditional mazes: railway mazes. These have junctions like those on railway tracks, and you have to take a route through them that a train could follow, one with no sharp turns. They are a good way to cram a complicated maze into a small space.

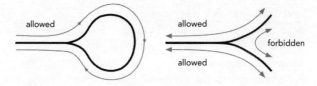

Allowed and forbidden routes at junctions

His son, the mathematician Roger Penrose, took the idea further. One of his mazes is carved in stone on the Luppitt Millennium Bench, in Devon, England. That one's a bit difficult, so here's a simpler example for you to tackle.

The map overleaf shows the rail network of Tardy Trains. The 10.33 train starts at station S and must finish at station F. The train cannot reverse direction by slowing down and then going backwards, but it can travel along a line in either direction if the track loops back on itself. At points, where two branches join, the train may take any smooth path. What route does the train take?

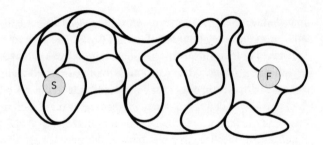

The maze

See page 252 for the answer, and further information including the Luppitt Millennium Maze.

. . . .

Soames Meets Watsup

A fine drizzle, of the kind that looks innocuous but quickly soaks you to the skin, was falling on the good citizens of London, and on the bad, as they scurried along Baker Street on errands admirable or nefarious, dodging the puddles. The not-so-famous detective was in his habitual position at the window, staring hopelessly into the gloom, grumbling to himself about his dire finances, and feeling depressed. His incisive solution to the Scandal of the Stolen Sovereign had brought in enough to get Mrs Soapsuds temporarily off his back, but now that the emotional rush of success had subsided, he felt lonely and unappreciated.

Perhaps he needed a like-minded companion? One who could share the daily cut-and-thrust of his personal vendetta against crime, and the intellectual challenge of unravelling the clues that its perpetrators scattered so carelessly across the landscape? But where could he find such a person? He had no idea where to start.

His black mood was interrupted by the appearance of a sturdy figure striding purposefully towards the premises opposite. Instinctively, Soames judged him to be a medical man, recently retired from the army. Well-dressed, well-heeled: yet another wealthy client for that overrated jackass Hol—

But no! The figure inspected the house number, shook his head, and spun on his heels. As he crossed the road, narrowly dodging a hansom cab, the brim of his hat concealed his face, but his body language showed determination, perhaps verging on desperation. Observing the man more closely, now that his interest had been piqued, Soames realised that his coat was not new, as he had first thought. It had been expertly repaired . . . in Old Compton Street, by the look of the stitching. On a Thursday, when the head seamstresses took a half-day off. *Down at heel, not well-heeled*, he corrected his initial impression, as the man disappeared from view, apparently heading for the doorway below.

A pause: then the bell rang.

Soames waited. A knock at the door announced his long-suffering landlady Mrs Soapsuds, clad in one of her habitual floral print dresses and wearing a large pinafore. "A gentleman to see you, Mr Soames," she sniffed. "Shall I show him up?"

Soames nodded, and Mrs Soapsuds slouched off down the stairs. A minute later she knocked again, and the medical gentleman entered. Soames waved at her to shut the door and return to her customary place behind the net curtains in her sitting-room on the ground floor, which she did with evident reluctance.

The gentleman listened for a moment, and suddenly tugged the door open, stepping back to allow Mrs Soapsuds to fall sideways to the floor.

"The—uh—mat. Needed dusting," she explained, picking herself up. Soames silently noted that his landlady also needed dusting, gave her a thin smile, and waved her away. Once more the door closed.

"My card," the man said.

Soames placed the visiting card face down, unread, and studied the new arrival from head to toe. After a few seconds he said, "Not much of note to identify you."

"Pardon?"

"Except the obvious, of course. You have been in

Al-Jebraistan for the last four years, serving as a surgeon with the Royal Sixth Dragoons. You narrowly escaped a serious wound at the Battle of Q'drat. Your period of service ended soon after, and you decided—after some soul-searching—to return to England, which you did early this year." Soames peered more closely, and added, "You keep four cats."

As the man's jaw dropped, Soames turned over the card. "Dr John Watsup," he read. "Surgeon, Royal Sixth Dragoons, retired." His face showed no emotion at this confirmation of his deductions, for it had been inevitable. "Please sit down, sir, and tell me of the crime that has been committed against you. I can assure you that—"

Watsup laughed, a friendly chuckle. "Mr Soames, I am delighted to have met you at last, for your fame has spread far and wide. Your deductions about my person prove that you fully deserve the acclaim that I have encountered. Your modesty at the feat becomes you. But I do not come primarily as a prospective client. Rather, I am seeking a position in your employ. Medicine no longer appeals to me—nor would it to you if you had seen the sights I have been forced to endure at the battlefront. But I am a man of action, I continue to crave excitement, I still have my service revolver, and . . . by the way, how did you *do* that?"

Soames, ignoring a growing feeling that he was being mistaken for the inhabitant of number 221B, sat down facing Watsup. "By your bearing, sir, I had you marked as a military man before you crossed the road. My eyesight is preternaturally keen, and you have the hands of a surgeon, strong yet lacking the ingrained stains of manual labour. Last December the *Times* reported that the four-year campaign in Al-Jebraistan was coming to a close and the Royal Sixth Dragoons were returning to England after fighting a decisive but costly battle at Q'drat. You are wearing the appropriate regimental boots, and the wear-patterns on them show you have been back in England for some time. You have a slight scar along your jawbone, almost healed, which was obviously caused by a musket-ball of non-European design—I have written a brief monograph on firearm injuries in the Far East, I must read it to you some time. You are a man of action, as evinced by the way you handled Mrs Soapsuds's snooping, so you would not have retired from military service voluntarily. If you had been given a dishonourable discharge I would have seen it reported in the scandal sheets, but nothing of the kind has been published recently. Your coat bears four different types of cat hair—not just four colours, which might indicate a single tabby, but different lengths and textures... I will spare you a list of their breeds."

"Astonishing!"

"To be candid, I must also admit that your face is familiar. I am sure that somewhere—ah, yes! I have it! A small article in last week's *Chronicle*, with a photograph... Dr John Watsup, originator of the well-known phrase 'Watsup, doc?' Your fame exceeds my own, Doctor."

"You are too kind, Mr Soames."

"No, merely realistic. But if we are to work together, you must convince me that you can *think* as well as act. Let us see." And Soames wrote the digits

4   9

on the back of an envelope. "I want you to insert one standard arithmetical symbol, to produce a whole number between 1 and 9."

Watsup pursed his lips in concentration. "A plus sign... no, 13 is too large. A minus—no the result is negative. Neither multiplication nor division will do. Of course! A square root! Oh, no: $4\sqrt{9} = 12$, again too large." He scratched his head. "I am stumped. It is impossible."

"I assure you there is a solution."

The silence was broken only by the ticking of a clock on the mantelpiece. Suddenly, Watsup's face lit up. "I have it!" He picked up the envelope, added a single symbol, and handed it to Soames.

"You pass the first test, Doctor."

What did Watsup write? See page 252 for the answer.

Geomagic Squares

A magic square is made from numbers, which give the same total along any row, column, or diagonal. Lee Sallows has invented a geometric analogue, the geomagic square. This is a square array of shapes, such that the shapes in any row, column, or diagonal fit together like a jigsaw to make the same overall shape. The pieces can be rotated or reflected if necessary. The left-hand figure shows how this goes; the right-hand one is a puzzle for you to solve. *See page 253 for the answer.*

Sallows has invented many other geomagic squares, along with generalisations such as a geomagic triangle. See *The Mathematical Intelligencer* 33 No. 4 (2011) 25–31 and his website

http://www.GeomagicSquares.com/

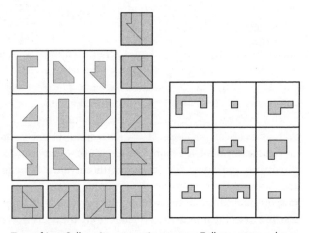

Two of Lee Sallows's geomagic squares. Follow a row, column, or diagonal to find the assembled jigsaw using the corresponding pieces. *Left*: A completed example. *Right*: Your task is find the assembled jigsaws for all rows, columns, and diagonals.

What Shape is an Orange Peel?

There are many ways to peel an orange. Some of us break bits off.
Some try to remove the entire peel in a single irregular lump.
This usually produces several lumps and a lot of juice. Others are
more systematic, peeling oranges very carefully with a knife,
making a spiral cut from the top of the orange down to the
bottom. I personally prefer a mess on the table and a quick
orange, but there you go.

Left: Peeling an orange with a knife. *Middle*: The peel laid out
flat. *Right*: Cornu spiral.

In 2012 Laurent Bartholdi and André Henriques wondered
what shape the peel would form if it were laid out flat. Using a
thin knife, and being careful to keep the peel the same width all
the way along, they obtained a beautiful double spiral. It
reminded them of a famous mathematical double spiral,
variously known as a Cornu spiral, Euler spiral, clothoid, or spiro.

This curve has been known since 1744, when Euler
discovered one of its basic properties. Its curvature ($1/r$ where r is
the radius of the best-fitting circle) at any given point is
proportional to the distance along the curve to that point,
starting from the middle. The further along the curve you go, the
more tightly it curves, which is why the spiral regions become
ever more closely wound. The physicist Marie Alfred Cornu came
across the same curve in the physics of light, when it diffracts at a
straight edge. Railway engineers have used the curve to provide a
smooth transition between a straight piece of track and a circular
one.

Bartholdi and Henriques proved that the resemblance
between orange peel and this shape is no accident. They wrote

down an equation describing the orange-peel shape for strips of any fixed width, and proved that when the width becomes as small as we please, the shape resembles the Cornu spiral ever more closely. They remarked that this spiral "has had many discoveries across history; ours occurred over breakfast."

See page 253 for further information.

· ·

How to Win the Lottery?

Please note the question mark.

To win the jackpot in the UK National Lottery (unimaginatively rebranded as 'Lotto') you have to choose six numbers from the range 1–49 that match those drawn on the day by a Lotto machine. There are other ways to win smaller prizes, but let's stick to that one. The balls are drawn in a random order, but the results are then converted to numerical order to make it easier to find out whether you've won. So a draw like

13   15   8   48   47   36

is reordered as

8   13   15   36   47   48

and in this case the smallest number is 8, the second smallest is 13, and so on.

Probability theory tells us that when all numbers are equally likely, as they should be, then within a chosen set of six:

The most likely smallest number is 1.
The most likely second smallest number is 10.
The most likely third smallest number is 20.
The most likely fourth smallest number is 30.
The most likely fifth smallest number is 40.
The most likely largest number is 49.

These statements are correct. The first is true because if 1 turns up, then it must be the smallest, no matter what else happens.

That's not the case for 2, however, because there is a small chance that 1 will turn up and sneak underneath it. This makes it slightly less likely that 2 will be the smallest after all six balls are drawn.

OK, those are mathematical facts. So it looks as though you can improve your chance of winning if you pick

    1   10   20   30   40   49

because each choice is the most likely number to occur in that position.

Is this correct? See page 253 for the answer.

• •

The Green Socks ~~Caper~~ Incident

"You have passed the first test, Doctor. But the real test will be to observe how you handle a criminal investigation."

"I am ready, Mr Soames. When shall we begin?"

"No time like the present."

"I agree, we are both men of action. Which case shall it be?"

"Your own."

"But—"

"Am I mistaken in thinking that although your reason for coming here was to seek employment, you have also been the victim of a crime?"

"No, but how—"

"When you first entered this room, I was instinctively aware that you were seeking my assistance. You were attempting to conceal it, but I saw it in your face and in your bearing. When I tested my deduction by speaking of 'the crime that has been committed against you', your reply was evasive. You stated that you had not come *primarily* as a prospective client."

Watsup sighed, slumping in his chair. "I was worried that mentioning my own case might have an adverse effect on your decision about engaging my services, by suggesting that I was merely seeking free advice. Once again you have seen through me, Mr Soames."

"That was inevitable. We may dispense with formalities. You may call me Soames. And I shall call you Watsup."

"An honour, Mr—er, Soames." Watsup, clearly upset, took a moment to steady himself. "It is a simple matter, of a kind that you will have encountered many times before."

"A burglary."

"Yes. How did—no matter. It happened earlier this year, and I immediately requested professional assistance from your neighbour across the road. After a month in which he made absolutely no progress, he declared that the matter was too trivial to interest his mighty talents, and showed me the door. Hearing by a fortunate accident of your own exploits, it occurred to me that you might succeed where the great luminary had failed."

It was clear to Watsup that he now had Soames's full attention.

"I vow to help you solve this crime, to prove my own worth to you," Watsup said with some emotion. "If we succeed—nay, *when* we succeed—my hopes of a more permanent engagement will be enhanced. I can pay you no fee, but I can offer my own unpaid services for two months. During which time I will ensure a steady flow of clients by singing your praises to the gentry, enough to keep us both fed and housed in moderate comfort."

"I confess that such an arrangement does have some appeal," said Soames. "I have been seeking what our transatlantic friends refer to as a 'sidekick' for some time. Your exposure of my landlady's nosiness gives me additional confidence that you will fit the bill admirably, but we shall see. Er—speaking of bills, you don't happen to have a five-pound note on you, by any chance? Mrs Soapsuds is always complaining about the unpaid rent... No, no, I see you are as strapped for cash as I. Together we shall overcome our mutual impecuniosity.

"Now, tell me of the crime."

"As I was saying, it is a simple matter," said Watsup. "My house was burgled, and my priceless collection of Al-Jebrian ceremonial daggers, representing the majority of my wealth, was stolen."

"Whence your present financial state."

"Indeed. I had planned to have them auctioned at Sotheby's."

"Were there any clues?"

"Just one. A green sock, left at the scene of the crime."

"What shade of green? What material? Cotton? Wool?"

"I do not know, Soames."

"These things matter, Watsup. Many a man has been hanged because of the precise colour of the dye in a single strand of darning wool. Or escaped the noose for lack of such evidence."

Watsup nodded, absorbing the lesson. "All the information I have was provided by the police."

"That explains its paucity, of course. Pray proceed."

"The police narrowed responsibility for the crime down to three men: George Green, Bill Brown, and Wally White."

Soames nodded thoughtfully. "The 'usual suspects', as I had already surmised. They operate in the Boswell Street area."

"How did you know I live in Boswell Street?" said Watsup in astonishment.

"Your address is on your card."

"Oh. In any case, one of those three was definitely the criminal. The police made enquiries, and found that each man habitually wore a jacket and trousers."

"Most men do, Watsup. Even the lower classes."

"Yes. But also, socks."

Soames pricked up his ears. "A feature perhaps of mild interest. It shows that these men have an income beyond their means."

"I'm sorry, Soames; I really don't see—"

"You have never encountered Messrs Brown, Green, and White."

"Ah."

"Please avoid distracting remarks, Watsup, and get to the point."

"Apparently it was each man's invariable habit to dress in garments whose colours were exactly the same on all occasions. Subtle traces at the scene of the crime—"

"Yes, yes," Soames muttered impatiently. "Threads adhering to the broken glass. Plain as the nose on a donkey."

"—uh, well, yes, as I was saying, threads. These indicated that the thief had used one of his socks to muffle the sound of breaking window-glass, and that the sock was green. Witnesses confirmed that between them the three men wore one jacket of each colour, one pair of trousers of each colour, and one matching pair of socks of each colour. None of them wore two or more garments of the same colour—counting a pair of socks as a single garment, you appreciate, since even ruffians such as these would not wear socks that do not match. That would be most improper."

"And did you deduce anything of consequence from that information?"

"Each of the suspects must have worn exactly one garment with the same colour as his name," said Watsup instantly. "If we deduce the colour, we find the criminal."

Soames leaned back in his chair. "Very good. Perhaps we will be able to work together. Anything else?"

"I came to the conclusion that the information to hand was insufficient to determine the criminal. The police eventually admitted as much, so I suggested they should seek further evidence."

"And did they find any?"

"After I had offered some more specific advice, they did." Watsup handed Soames a sheet of paper. "Part of the police report," he explained. The document read:

Extract from Report of Investigation by Constable J.K. Wuggins of the Holborn Division of the Metropolitan Police Force

1 Brown's socks were the same colour as White's jacket.

2 The person whose name was the colour of White's trousers

wore socks whose colour was not the name of the person wearing a white jacket.

3   The colour of the jacket of the person whose name is the colour of Green's socks is different from the colour of Brown's trousers.

"And there you have it," said Watsup. "If we can determine the thief, then the police will be able to obtain a search warrant. With luck they will find my missing daggers, which would amount to incontrovertible proof of guilt. But they are stumped, and your overrated neighbour is as baffled as I—which is why he pretends that the case has no interest."

Soames chuckled. "On the contrary, my dear Watsup. Thanks to your assiduous devotion to persuading the police to investigate the circumstances of the crime in sufficient depth, there is enough information to determine the guilty party. The deduction is of course elementary."

"How can you be so sure?"

"You will come to know my methods," said Soames enigmatically.

"Who is the criminal, then?"

"We will find out when we make the deduction."

Watsup produced a new, fat, currently blank, notebook, and wrote:

Memoirs
by Dr John Watsup (M. Chir., R.M.C.S., *retd*)
One: The Green Socks Caper

Soames, reading the words upside down, said quietly "This is not a penny dreadful, Watsup." Watsup crossed out 'Caper' and inserted 'Incident'. Then, pursing his lips, he began to record their joint analysis. With a few hiccups along the way, the identity of the thief soon emerged.

See page 254 for the answer.

"I shall send Inspector Roulade a telegram immediately," Soames declared. "He will send constables to raid the man's premises. No doubt they will find your daggers there, since the man we have identified is notoriously slow to fence stolen property. He likes to *gloat*, Watsup, an error that has put him behind bars more than once.

"And that wraps up our first case together!" His excitement quickly subsided as he added, "Your assistance was vital, but unfortunately the outcome of our deliberations does not improve our finances, since it is *your* case."

"There will be some improvement. I will regain my daggers."

"I fear the police will hold them as evidence until after the trial. Even so, we may consider it a harbinger of more profitable times to come, eh, Watsup?"

. . . .

Consecutive Cubes

The cubes of the three consecutive numbers 1, 2, 3 are 1, 8, 27, which add up to 36, a perfect square. What are the next three consecutive cubes whose sum is a square?

See page 258 for the answer.

. . . .

Adonis Asteroid Mousterian

AD	IN	SO
IS	DO	AN
NO	AS	ID

ADONIS

AS	IR	ED	TO
DO	ET	IS	RA
IT	AD	OR	ES
RE	SO	AT	ID

ASTEROID

EN	MA	IR	SO	UT
IS	TO	NU	ME	RA
MU	RE	AS	IT	NO
AT	IN	OM	UR	ES
OR	US	ET	AN	MI

MOUSTERIAN

Three of Farrell's magic word squares

Jeremiah Farrell published some amazing magic word squares in *The Journal of Recreational Linguistics* 33 (May 2000) 83–92. These are samples. The entries in each square are two-letter words appearing in standard dictionaries. The same letters appear in each row and column, and in each of the two main diagonals of the order 4 and 5 squares. Every row and column is an anagram (though not a meaningful one) of the same dictionary word, which is written underneath. *Mousterian* is a style of flint tool used by some Neanderthals, by the way.

You may feel that word arrangements are not terribly mathematical. However, puzzle buffs tend to enjoy both, and I am inclined to see word games as combinatorial problems posed with irregular constraints; namely, the dictionary. But these squares have mathematical features too. If numbers are suitably assigned to letters, and the numbers corresponding to each pair of letters in a given square are added together, the resulting numerical square is also magic. That is, the numbers in every row, column, and (except for the 3×3 square) diagonal add to the same amount.

Of course, this property holds for any assignment of numbers, except on the diagonals of the 3×3 square, because each letter occurs exactly once in each row, column, and (except for the 3×3 square) diagonal. However, with the correct choice, the numbers run from 0–8, 0–15, and 0–24 respectively. The assignments are different for each magic word square.

Which numbers correspond to which letter? See page 258 for the answer.

. . . .

Two Square Quickies

1   What is the largest perfect square that uses each digit 123456789 exactly once?

2   What is the smallest perfect square that uses each digit 123456789 exactly once?

See page 259 for the answers.

. . . .

Caught Clean-Handed

John Napier

John Napier, eighth Laird of Merchistoun (now Merchiston, part of Edinburgh), is famous for inventing logarithms in 1614. But there was a darker side to his nature: he dabbled in alchemy and necromancy. He was widely believed to be a magician, and his 'familiar', or magical companion, was a black cockerel.

He used it to catch servants who were stealing. He would lock the suspect in a room with the cockerel and tell them to stroke it, saying that his magical bird would unerringly detect the guilty. All very mystical—but Napier knew exactly what he was doing. He coated the cockerel with a thin layer of soot. An innocent servant would stroke the bird as instructed, and get soot on their hands. A guilty one, fearing detection, would avoid stroking the bird.

Clean hands proved you were guilty.

The Adventure of the Cardboard Boxes*

From the Memoirs of Dr Watsup

With the restoration of my valuable ceremonial daggers, and our partnership's growing reputation for cracking the uncrackable, solving the insoluble, and unscrewing the inscrutable, our personal finances were improving by the day. The elite of England were virtually queuing up to engage our services, and my notebooks contain many of my friend's successes: the Mystery of the Missing Mountain, the Vaporised Viscount, and the Bald-Headed League. None of these cases, however, captures Soames's talents in their purest form: his ability to discern significant aspects of apparently ordinary objects and events that few others would notice. The Giant Bat of St Albans naturally springs to mind, but the ramifications of the case are too arcane and complex to describe here.

The curious events of Christmas 18—, however, suit my purpose admirably, and deserve wider appreciation. (I am forced to conceal the exact date and most of what happened to avoid embarrassment to a famous operatic contralto and several Cabinet Ministers.)

I was seated at my writing-desk, recording details of Soames's most recent cases, while he carried out a seemingly endless series of experiments with my old service revolver and vases of chrysanthemums. Our separate activities were interrupted by Mrs Soapsuds, who deposited two cardboard boxes of different sizes, each tied with ribbons. "Some Christmas presents for you, Mr Soames," she announced.

Soames eyed the packages. They bore his address and some cancelled postage stamps with illegible postmarks. They were rectangular in shape ... well, technically a rectangle is two-

---

* This and all subsequent cases investigated by Soames and Watsup are reproduced (slightly edited) from *The Memoirs of Dr Watsup: Being a Personal Account of the Unsung Genius of an Underrated Private Detective*, Bromley, Thrackle & Sons, Manchester 1897.

dimensional, so they were actually rectangular parallelepipeds. Cuboids.

Box-shaped.

He took out a ruler and measured their dimensions. "Remarkable," he muttered. "And very, very disturbing."

I have learned to respect such judgements, however peculiar they first may seem. I stopped thinking of the packages as Christmas presents, tried to dismiss a growing suspicion that they were bombs, and did my best to *observe*. Finally, I realised that they had been tied with rather more ribbon than was strictly necessary.

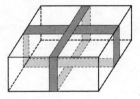 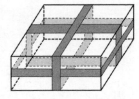

Left: Watsup's usual arrangement of ribbons. *Right*: The way each package was tied.

"The ribbons form a cross on every face of the packages," I said. "When I wrap parcels, I normally tie the ribbon so that it forms a cross on the top and bottom faces, and runs vertically down each of the other four."

"You do indeed."

Clearly my analysis was incomplete. I racked my brains. "Umm—there is no bow."

"Correct, Watsup."

Still incomplete. I scratched my head. "That is all I can observe."

"That is all you can *see*, Watsup. You have noticed everything *except* the crucial pattern. I fear that terrible deeds are afoot."

I confessed that I saw nothing terrible in two Christmas presents. Then a thought struck me. "Do you mean that the boxes contain severed body parts, Soames?"

He laughed. "No, they are *almost* empty," he said, picking

them up and shaking them. "But surely you realise that this particular type of ribbon can be bought only from the Ladies Wilberforce?"

"Regrettably, no, but I bow to your superior knowledge. The establishment, however, is familiar to me. It is a haberdashery in Eastcastle Street." The penny dropped. "Soames! That is where that terrible murder was committed! It was—"

"—in all the papers. Yes, Watsup."

"The evidence was convincing, but the body has not yet been found."

Soames nodded, his face grim. "It will be."

"When?"

"Shortly after I open these boxes."

He put on a pair of gloves and set to work unwrapping them. "Undoubtedly this is the work of the Cartonari, Watsup." When I stared at him blankly, he added: "an Italian secret society. But it is better that you remain ignorant." Despite all my entreaties, he refused to say more.

He opened both boxes. "As I suspected. One is empty, but the other contains *this*." He held up a small rectangle of paper.

"What is it?"

He passed it to me. "A left-luggage ticket," said I. "It must be a message from the murderer. But the serial number has been torn off, and so has the name of the station."

"Only to be expected, Watsup. He—for by the footprints in the blood the criminal was assuredly a man, and a large one at that—is taunting us. But we shall have the better of him. The station is of course obvious from the arrangement of the ribbon."

"Uh—pardon?"

"Together with the value of the stamps, which eliminates the possibility of Charing Cross."

This made little sense, so I picked up a package and counted five one-shilling stamps. "An absurd amount to pay for an empty package." I said, puzzled.

"Not if you wish to send a message. What is another name for five shillings?"

"One crown."

"And a crown is symbolic of?"

"Our own dear queen."

"Close, Watsup, but you have failed to take account of the shape of the ribbon."

"It is a cross."

"So the stamps indicate 'king', not queen. The station is *King's Cross*, man! But there is more. Answer me this, Watsup. Why did the criminal send me two large boxes when one was empty? One small envelope would have sufficed to send a ticket."

After a long silence I shook my head. "I have no idea."

"There must be something significant about the relationship between the boxes. And indeed there is, as I realised as soon as I measured their sizes." He handed me the ruler. "Use this."

I repeated his measurements. "The length, breadth, and height of each package is a whole number of inches," I said. "Otherwise no pattern springs to mind."

He sighed. "You did not observe the strange coincidence?"

"What strange coincidence?"

"Both packages have the same volume, and they both use the same total length of ribbon. Indeed, their measurements are the smallest nonzero whole numbers with that property."

"Which leads you to conclude—oh, of course! The volume and length together give the serial number on the ticket. There are two distinct ways to string them together, of course, but we can easily check both."

Soames shook his head. "No, no. The murderer would have needed an accomplice at the ticket office to arrange that, even if a ticket with that number existed. It is much simpler: he has marked some item of left luggage with these two numbers. Inside, there will be something that tells us where to find it."

"To find what?"

"Is it not obvious? The body."

"I take my hat off to you, Soames," I said. "Or would, were I wearing one. But will finding the body lead us to the murderer?"

"It will be useful evidence, but inconclusive. However, there is more to be gleaned. Sometimes a criminal believes himself so clever that he deliberately leaves clues, certain that the investigating authorities will be too stupid to notice them. The Cartonari are an arrogant bunch, and this is typical of them. Now, there is a natural question that leads on from the remarkable arithmetic of these boxes. What is the smallest set of *three* boxes with a similar property?"

His train of thought became instantly apparent to me. "You expect to receive such boxes in the near future! With another torn ticket! So there is to be another murder, eh?" I started looking for my revolver. "We must stop it!"

"I fear that it has already been committed, but with good fortune we may perhaps prevent a third death. Tonight the murderer will be depositing some item—it might be anything—as left luggage at one of the main London railway stations. Then he will send us the boxes. If we can work out the numbers ahead of time, we can alert Inspector Roulade. He will send police to all of the mainline stations. They cannot check every passenger who deposits luggage, for that would alert the criminal, but they can keep a watch for anyone depositing an item with those three numbers marked on it, and arrest them. Inside will be the location of the second body. When it is found, the evidence of guilt will be overwhelming."

In the event, it was not quite that simple, and Soames and I had to intervene after the police let the man slip. Fortunately the three packages, which duly arrived next day in the afternoon post, provided new clues, and we discovered that the murder was part of a far more extensive plot. The tortuous paths down which our deductions led us, and the blood-curdling secrets that we unearthed—I speak literally—can never be made public, as I have explained. But we did eventually catch the criminal. And Soames has permitted me to reveal the answers to the two questions that were central to the entire investigation.

What are the dimensions of the two boxes? What are they for three boxes? See page 259 for the answer.

The RATS Sequence

1, 2, 4, 8, 16, . . . What comes next? It's tempting to leap to conclusions and plump for 32. But suppose I tell you that the sequence I have in mind actually goes

$$1 \quad 2 \quad 4 \quad 8 \quad 16 \quad 77 \quad 145 \quad 668$$

Now what comes next? There isn't a unique answer, of course: by finding sufficiently contrived rules you can fit a formula to any finite sequence. In *Mathematics Made Difficult* Carl Linderholm has an entire chapter explaining why you can always answer 'what comes next in this sequence' with '19'. But, bear with me: there is a simple rule for the sequence above. The title of this section is a clue, but a much too obscure one to be of any use to you.

See page 260 for the answer.

Birthdays Are Good for You

Statistics show that people who have most birthdays live the longest.

Larry Lorenzoni

Mathematical Dates

In recent years numerous dates have been associated with aspects of mathematics, based on numerical similarities, leading to that date being declared a special day. No one attaches any significance to such days, aside from the numerical similarity. They do not foretell the End of the World or anything of that kind—as far as we know. Nothing special happens on them aside from mathematical celebrations and occasional comments in the media. But they're fun—and an excuse to interest the media in more significant mathematics. Or at least to mention the word.

Among them are the following. Many are associated with alternative dates, because in the American system for dates, the month precedes the day. In the British system, the day comes first. A certain amount of calendric licence is allowed, such as omitting zeros.

Pi Day

14 March, American system of dates: 3/14. [π ~ 3.14.] Quasi-official day since 1988 in San Francisco. Recognised by a non-binding resolution in the US House of Representatives.

Pi Minute

1.59 on March 14 (American system). 3/14 1:59. [π ~ 3.14159.]
    More accurately still: 1.59 and 26 seconds. 3/14 1:59:26. [π ~ 3.1415926.]

Pi Approximation Day

22 July, British system of dates: 22/7. [π ~ 22/7.]

123456789 Day

Sorry, you missed it. This once-in-a-lifetime event occurred on 7 August 2009 (British system), or 8 July 2009 (American system), shortly after 12.34 p.m. The date and time were: 12:34:56 7/8/(0)9. But some of you might witness 1234567890 Day in 2090.

Onesday

You missed this too. It occurred on 11 November 2011 (either system) at 11 minutes, 11 seconds past 11 o'clock. The date and time were: 11:11:11 11/11/11.

Twosday

As I write, this will happen a few years from now. This is your opportunity! 2 February 2022: 22:22:22 2/2/22.

Palindrome Day

A palindrome reads the same in reverse—as in 'evil rats on no star live'. 20 February 2002 at 8.02 p.m. (British system, 24-hour clock): 20:02 20/02/2002.

    This repeats the *same* palindrome three times. When will the next such day be, on the British system? What was the next date after the one above, again using the British system, on which the whole thing was palindromic?

For the answers see page 261.

Fibonacci Day (Short version)
3 May 2008 (British system), 5 March 2008 (American system): 3/5/(0)8.

Fibonacci Day (Long version)
5 August 2013 (British system), 2 minutes 3 seconds past 1 o'clock on 8 May 2013 American system): 1:2:3 5/8/13.

Primes Day
2 March 2011 (British system), 3 February 2011 (American system): 2:3 5/7/11.

. . . .

The Hound of the Basketballs

From the Memoirs of Dr Watsup

"A lady to see you, Mr Soames," said Mrs Soapsuds.

Soames and I sprang to our feet. A woman of indeterminate age entered—indeterminate because she wore a dark veil.

"There is no need to disguise yourself, Lady Hyacinth," said Soames.

With a gasp, she pulled off her veil. "How—"

"The extraordinary events at Basket Hall have been in the headlines for a week," said Soames. "I have been following the case closely, and my competitor across the road has made no progress. It was only a matter of time before you sought my services. Also, I recognised your driver's hat, which is quite unique among the servants of the aristocracy."

"*Basquet*, not Basket," said Lady Hyacinth with a sniff, making the word sound French.

"Come now, madam," said Soames. "The house has been in

the Basket family for seven generations, ever since Honoria Thumpingham-Maddely married the Third Earl."

"Well, yes, but that was *then*. The spelling and pronunciation have . . . er—"

"Been modernised," I interjected, hoping to smooth increasingly troubled waters of mutual disaffection. At the same time I gave Soames a sharp look, unobserved by Her Ladyship. To his credit, he took my point.

"It was a gigantic black hound!" she suddenly cried, the words sounding as if they were being torn from her throat. "With great slavering jaws that dripped blood!"

"You saw it?"

"Well, no, but the boy who looks after the piglets . . . Nicky, that's the name. Or is it Ricky? Anyway, he said he caught a glimpse of the vile horror just as it disappeared."

"In the dark," Soames pointed out. "From a distance of one hundred and seventy yards. Michael Jenkins is short-sighted. But no matter, the evidence will lead us to the truth eventually. I gather that the animal did no harm to any human being?"

"Well, no," she agreed. "Not directly, though my poor husband . . . You see, the hound ruined a tradition that goes back even to before the Third Earl of Bask—Basquet."

I belatedly remembered my etiquette. "Dr John Watsup, at your service, madam. I regret that unlike my companion I have not been following the news. If you would be so kind as to enlighten me?"

"Ah. Yes. Um." She gathered her skirts and her wits. "It was a few nights before Midwinter's Eve, and my husband Edmund— that's Lord Basquet, of course—had arranged twelve antique stone spheres—"

"Known for centuries as basketballs," Soames interrupted.

"Well, yes, but we can't modernise *everything*, Mr Soames. There are traditions. Anyway, my husband had arranged the balls on the lawn of our stately home in an age-old family symbol. Only the heir to the male line knows exactly what the symbol is, and no one else is permitted to observe the ceremony, but over the years it has become common knowledge that it consists of seven straight rows of balls, with four in each row.

"Edmund was rehearsing for a ceremony which must be carried out without fail every Midwinter's Eve. But when we awoke the following morning, we were horrified to find that some of the balls had been moved!"

"But you have said that no one save Lord Basquet is permitted to see the arrangement," Soames objected.

"The circumstances were exceptional. His Lordship went to recover the balls, but failed to return. Eventually one of the serving-maids was sent to find him—Lavinia is blind, you see, but very capable. She came back in tears screaming that his Lordship was lying on the ground, not moving. Fearing he was dead, the rest of us disobeyed the time-honoured injunction and rushed to the scene. I was just in time to hear Edmund gasp 'Moved!' Then he was still. He has been in a stupor ever since, Mr Soames. It is most distressing."

"*Moved*," said I. "In what manner, madam?"

"No longer in the same place, Dr Watsup."

"I mean, moved *where*?"

"They now formed a star, Dr Watsup."

"Yes! One having only six straight rows with four balls in each row," said Soames, sketching rapidly on a sheet of paper. "This fact has been widely reported and has the ring of truth, for

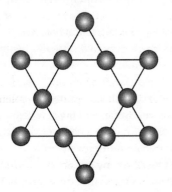

The balls after the hound had moved them

it is unlikely to have been invented, being too mentally taxing for the yellow press. It also proves that we could have *deduced* that the balls had been moved without having to rely on his Lordship's last gasp—"

"Last gasp *to date*," I said hurriedly, before Soames triggered a fresh wave of wails.

"Could you not move them back?" I enquired, when Her Ladyship had regained some of her composure.

"Nay!" she cried. I have long remarked that the English aristocracy have a noticeable tendency to resemble horses.

"Why not?"

"As I told you, only his Lordship knows the exact arrangement that tradition requires, and the doctors say he may never recover!"

"Were there not marks where the balls had originally rested?"

"Perhaps, but they were obscured by the tracks of the terrible hound!"

"I will bring my most powerful magnifying glass, then," said Soames, with a straight face. A thought must have struck him, because he suddenly froze. "You said 'must'."

"I did? When?"

"A few minutes ago you said that the ceremony *must* be carried out without fail every year. It has just occurred to me that your choice of words may be significant. Explain them."

"According to an ancient Transylvanian prophecy, if the twelve basketballs are not deployed correctly on Midwinter's Eve, the House of Bask—er, Basquet—will fall, and be utterly destroyed! And we have only three days to do it! Oh, woe!" She began to sob.

"Calm yourself, madam," said I, wafting an open bottle of smelling-salts beneath her nose. "Please accept my condolences for His Lordship's unfortunate condition, and my assurances as a medical man that there is some faint chance, be it ever so slight, that it may undergo some miraculous improvement in due course." I have long prided myself on my impeccable bedside manner, which puts that of my friend Soames to shame, but on this occasion, unaccountably, Her Ladyship's response was to redouble her sobbing.

Soames paced the room, his face drawn. "Is it only the *form* of the arrangement that matters, Your Ladyship? Or does the orientation make a significant difference?"

"I beg your pardon?" said she, shaking her head as if to clear her mind.

"If the arrangement were correct *except for a rotation*, without changing the relative positions of the balls, would that be expected to trigger the dire events predicted?" Soames clarified.

Lady Basquet paused, considering the matter. "No. Definitely not. I recall Willy Willikins—that's the head gardener—suggesting to my husband that from time to time he might care to point the arrangement in a different direction, to avoid damaging the lawn. Edmund raised no objection."

"That is excellent news!" said Soames.

"Yes. Excellent," I echoed, having not the faintest idea why my detective friend was so pleased. Or, for that matter, what his question meant.

"Were there any signs of human intervention?" Soames asked.

"No. The head gardener swore that no human being save Edmund had set foot on the lawn. Young Dicky—"

"Micky."

"Vicky saw the terrible hound, but even he obtained only a fleeting glimpse as it leaped over the perimeter of the walled garden. We have some delightful peonies, Mr Soames, although they are not in bloom at this time of—"

"I will take the case," said Soames. "If Your Ladyship cares to return to Basquet Hall, I and my colleague will arrive on Thursday by the earliest possible slow train."

"No sooner, Mr Soames? But Thursday *is* Midwinter's Eve! The balls must be placed correctly before the sun sets!"

"I regret that I am detained until then by a small matter involving three Eastern potentates, six hundred thousand armed warriors, two disputed borders, and a stolen casket of emeralds and sapphires belonging to an obscure and arcane ancient religious order. And a flattened copper thimble, which I believe holds the key to the entire affair. However, I assure you that I am confident that your case will be settled satisfactorily before sunset on Thursday."

Her protestations notwithstanding, Soames remained adamant, and eventually Lady Hyacinth Basquet departed, snuffling quietly into a lace kerchief.

After she had left, I asked to which matter Soames had been referring, for I had heard nothing of this case. "A small fabrication on my part, Watsup," he confessed. "I have tickets for the opera this evening."

We arrived mid-afternoon on the Thursday, to be met at the station by a groom driving a governess cart. Or possibly a governess in a groom cart, my notes being illegible on this point. Lord Basquet remained in a coma, we were informed. Within a bare half hour we had arrived at the Hall, and Soames was inspecting the extensive lawns using an unusually large magnifying glass, a hairbrush, and a protractor.

"A chance for you to exercise your deductive skills, Watsup," he said.

"I see some disturbance of the grass, Soames."

"Correct Watsup. The tracks are highly complex, but they are mainly superimposed paw-prints of—" he lowered his voice so that I alone could hear—"a miniature poodle." In his normal voice he continued: "I am unable to discern the original placing of the balls, but unless I am very much mistaken—and I never am—it is clear that the animal moved precisely *four* of the balls."

"Is that significant, Mr Soames?" Lady Basquet asked nervously, cradling a miniature poodle in her arms.

Soames glanced my way.

"It is . . . possible . . . " I began, and saw Soames nod imperceptibly. Well, not *totally* imperceptibly, you appreciate, since if it *had* been imperceptible I would not have seen it. Taking this as veiled encouragement, I hazarded a guess: " . . . that this condition will make it possible to deduce the original layout."

"And does it?" she enquired, a hopeful look on her face.

What was the original arrangement of the basketballs? See page 261 for the answer.

. . . .

Digital Cubes

This is an oldie, but it opens up a less familiar question. The number 153 is equal to the sum of the cubes of its digits:

$$1^3 + 5^3 + 3^3 = 1 + 125 + 27 = 153$$

There are three other 3-digit numbers with the same property, excluding numbers like 001 with a leading zero. Can you find them?

For the answer see page 262.

. . . .

Narcissistic Numbers

The cubes puzzle has some notoriety, because in 1940 the famous pure mathematician Godfrey Harold Hardy wrote, in *A Mathematician's Apology*, that such puzzles have no mathematical merit because they depend on the notation employed (decimal) and are little more than coincidences. However, you can learn quite a lot of useful mathematics by trying to solve them, and generalisations (for example, to number bases other than 10) avoid the notational issue.

One of the offshoots of this puzzle is the concept of a *narcissistic number*, which is defined to be a number that is equal to the sum of the *n*th powers of its decimal digits for some *n*. The term *n-narcissistic number* is used if we want to make *n* explicit.

Fourth Powers of Digits (4-Narcissistic Numbers)

Write [abcd] for the number with digits a, b, c, d to distinguish this from the product abcd. That is, [abcd] = $1000a + 100b + 10c + d$. We must solve

$$[abcd] = a^4 + b^4 + c^4 + d^4$$

where all unknowns lie between 0 and 9. This is by no means a trivial task. Try it!

See page 263 for the answer.

Fifth Powers of Digits (5-Narcissistic Numbers)

This time the problem is to solve

$$[abcde] = a^5 + b^5 + c^5 + d^5 + e^5$$

and as you'd expect, that's even harder.

See page 263 for the answer.

Higher Powers of Digits (n-Narcissistic Numbers, $n \geq 6$)

It is easy to prove that n-narcissistic numbers exist only for $n \leq 60$, because whenever $n > 60$ we have $n.9^n < 10^{n-1}$. In 1985 Dik Winter proved that there are exactly 88 narcissistic numbers with nonzero first digit. For $n = 1$ they are the ten digits (we include 0 because it is the *only* digit in this case). For $n = 2$ they don't exist. For $n = 3$, 4, 5 see the answers to Digital Cubes (page 262) and the above two problems. For $n \geq 6$ they are:

n	n-narcissistic numbers
6	548834
7	1741725, 4210818, 9800817, 9926315
8	24678050, 24678051, 88593477
9	146511208, 472335975, 534494836, 912985153
10	4679307774
11	32164049650, 32164049651, 40028394225, 42678290603, 44708635679, 49388550606, 82693916578, 94204591914
14	28116440335967
16	4338281769391370, 4338281769391371

17 21897142587612075, 35641594208964132,
   35875699062250035
19 1517841543307505039, 3289582984443187032,
   4498128791164624869, 4929273885928088826
20 63105425988599693916
21 128468643043731391252, 449177399146038697307
23 21887696841122916288858, 2787969489305407447 1405,
   2790786500997705256781 4, 28361281321319229463398,
   35452590104031691935943
24 17408800593806529302 3722, 18845148544789789603 6875,
   23931366443004156935009 3
25 15504753342145015390888 94, 15532421628937718506 69378,
   37069079959554759886443 80, 37069079959554759886443 81,
   44220951180958996194 57938
27 121204998563613372405438 066,
   121270696006801314328439 376,
   128851796696487777842012 787,
   174650464499531377631639 254,
   177265453171792792366489 765
29 14607640612971980372614873 089,
   19008174136254279995012734 740,
   19008174136254279995012734 741,
   23866716435523975980390369 295
31 11450372757654910259242920 50346,
   19278904571429606975806362 36639,
   23090926826161903075096953 38915
32 17333509997782249308725103962772
33 18670996100153879010063413297 6990,
   18670996100153879010063413297 6991
34 11227632853293725415928229002 04593
35 12639369517103790328947807201 478392,
   12679937780272278566303885594 196922
37 12191672196254341215697358036 09966019
38 12815792078366059955099770545296129367
39 11513221901876399256509559779739 71522400,
   11513221901876399256509559779739 71522401

Piphilology, Piems, and Pilish

Now, I wish I could recollect pi.

'Eureka!' cried the great inventor.
Christmas pudding; Christmas pie
Is the problem's very centre.

See, I have a rhyme assisting
my feeble brain,
its tasks sometimes resisting.

How I wish I could enumerate pi easily, since all these horrible mnemonics prevent recalling any of pi's sequence more simply.

The last one gives the game away: these are mnemonics—memory aids—for π. There's even a word for this: *piphilology*. Count the letters in successive words: 3, 1, 4, 1, 5, . . .

Some of the many mnemonics for π were discussed in *Cabinet of Mathematical Curiosities*; here we recall one (the French mnemonic below) and look at a few more. There are hundreds, in many languages: see

http://en.wikipedia.org/wiki/Piphilology

http://uzweb.uz.ac.zw/science/maths/zimaths/pimnem.htm

One of the most famous is the French alexandrine (a poetic metre), which begins:

Que j'aime à faire apprendre
Un nombre utile aux sages!
Glorieux Archimède, artiste ingenieux,
Toi, de qui Syracuse loue encore le mérite!

and goes on until it gets to 126 places. I especially recommend the Portuguese:

Sou o medo e temor constante do menino vadio.

(I am the constant fear and terror of lazy boys.)

The Romanian:

Asa e bine a scrie renumitul si utilul numar.

(That's the way to write the famous and useful number.)

has the merit of directness and simplicity.

The poems are known as *piems*. The 32nd decimal place of π is 0, and a word of length 0 is invisible. However, there are ways round this obstacle. In *Pilish*, the cryptographic system normally used for π mnemonics, a ten-letter word counts as 0. Mike Keith's 'A Self-Referential Story', which encodes 402 digits of π [*The Mathematical Intelligencer* 8 (No.4) (1986) 56–57] employed a different set of rules. The longest examples I know of (to date, *Guinness Book of World Records* here we come, if I know my readers) are the short story 'Cadaeic cadenza' (3,834 digits) and the book *Not A Wake* (10,000 digits), also by Keith. The book begins:

> Now I fall, a tired suburbian in liquid under the trees
> Drifting alongside forests simmering red in the twilight
> over Europe.
> So scream with the old mischief, ask me another
> conundrum
> About bitterness of possible fortunes near a landscape
> Italian.
> A little happiness may sometimes intervene but usually
> fades.
> A missionary cries, striving to understand worthless,
> tedious life.
> Monotony's lost amid ocean movements
> As the bewildered sailors hesitate. I become salt,
> Submerging people in dazzling oceans of enshrouded
> unbelief.
> Christmas ornaments conspire.
> Beauty is, somewhat inevitably now, both
> Feelings of faith and eyes of rationalism.

Here ten-letter words count as 0, and longer ones count as two digits; for example, a 13-letter word counts as 13.

For a wealth of related information and other examples see Keith's website

http://cadaeic.net

• •

Clueless! 🔍

From the Memoirs of Dr Watsup

As I flick through the well-thumbed pages of my notebooks, my memories are drawn to innumerable mysteries that Soames solved by observing clues so subtle as to elude lesser minds, such as the Adventure of the Sussex Umpire (a remarkable locker-room mystery whose pivotal feature was a prematurely scuffed cricket ball), the Cow with the Crumpled Horn, the Attempted Triple Murder of the Diminutive Swine, and the Affair of the Missing Tart. But among them one stands out: a mystery whose sole clue was the absence of any clues whatsoever.

It was a damp, gloomy Tuesday, and the streets of central London were thick with smoke and fog. We had abandoned active pursuit of the perpetrators of crime for a period of introspection before a warm fire, accompanied by ample glasses of an amusingly presumptuous claret.

"I say, Soames," I remarked.

My colleague was searching through a thick stack of photographic plates of hoofprints in mud, produced using Eastman's new improvement of Maddox's gelatine process. His response was an irritated "Have you seen my collection of carthorse photographs anywhere, Watsup?", but I continued doggedly.

"This puzzle hasn't got a clue, Soames."

"It's not the only one," he muttered darkly.

"No, I mean—it hasn't got *any* clues."

I could see I now had his attention. He took the newspaper from my outstretched hand, and glanced at the diagram.

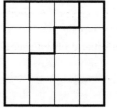

Clueless puzzle

"The unstated rules are obvious, Watsup."

"Why?"

"They must be simple enough to motivate the would-be solver, yet lead to a sufficiently challenging problem to retain their interest."

"No doubt. So what *are* the rules, Soames?"

"Clearly each row and each column must contain each number 1, 2, 3, 4 once only."

"Ah. It is a combinatorial puzzle, a type of Latin square."

"Yes, but there is more. The two regions outlined in thick black lines are obviously important. I conjecture that the numbers in each region must have the same total . . . Yes, that leads to a unique solution."

"Oh. I wonder what the answer is."

"You know my methods, Watsup. Use them." And he returned to his photographic plates.

See page 263 for the answer. For more clueless puzzles, see page 74.

. . . .

A Brief History of Sudoku

Modern readers will recognise Watsup's puzzle as a variant of sudoku. (If you've just come back from a forty-year trip to Proxima Centauri, this is a 9×9 array divided into nine 3×3 blocks, with some numbers inserted. You have to fill in the rest so that every row, every column, and every block contains each digit 1–9.)

Similar but significantly different puzzles have a long history, which goes right back to the Chinese *Lo Shu*, a magic square allegedly seen on the back of a turtle around 2100 BC. Jacques Ozanam's 1725 *Récréations Mathématiques et Physiques* included a puzzle about playing cards that gets slightly closer to sudoku. Take the 16 court cards (ace, king, queen, jack) and arrange them in a square so that each row and column contains all four face

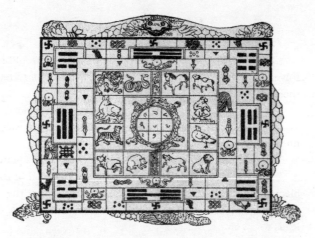

Lo Shu, centre, on the back of a small turtle, surrounded by the Chinese zodiac and the divinatory trigrams of the *I Ching*, all carried by a larger turtle, which according to myth first revealed the trigrams

values and all four suits. Kathleen Ollerenshaw showed that there are 1152 solutions, which reduce to just two basically different ones if we consider two solutions to be the same if one can be obtained from the other by permuting the face values and suits. There are $24 \times 24 = 576$ ways to do this to any given solution, and $1152/576 = 2$.

Can you find these two basically different solutions? See page 265 for the answer.

In 1782 Euler wrote about the thirty-six officers problem: can six regiments, each having six officers of different ranks, be arranged in a 6×6 square so that each row and column contains every rank and every regiment? Such arrangements were called Graeco-Latin squares because Latin letters A, B, C,... and Greek letters α, $\beta \gamma$, ... could be used to denote the ranks and regiments. He found methods for constructing Graeco-Latin squares whose order (the size of the square) is odd or doubly even: a multiple of four.

Aα	Bδ	Cβ	Dε	Eγ
Bβ	Cε	Dγ	Eα	Aδ
Cγ	Dα	Eδ	Aβ	Bε
Dδ	Eβ	Aε	Bγ	Cα
Eε	Aγ	Bα	Cδ	Dβ

An order 5 Graeco-Latin square

Euler conjectured that no such squares exist when the order is doubly odd: twice an odd number. This is obvious for order 2, and Gaston Tarry proved it in 1901 for order 6. However, in 1959 Raj Chandra Bose and Sharadchandra Shankar Shrikhande used a computer to find a Graeco-Latin square of order 22, and Ernest Parker found one of order 10. The three of them then proved that Euler's conjecture is false for all doubly odd orders greater than or equal to 10.

Square $n \times n$ arrays such that each row and column contains all of the numbers $1-n$ (each necessarily appearing once) became known as Latin squares, and Graeco-Latin squares were renamed orthogonal Latin squares. These topics are part of the branch of mathematics known as combinatorics, and they have applications to experimental design, the scheduling of competitions, and communications.

A completed sudoku grid is a Latin square, but there are extra conditions on the 3×3 blocks. In 1892 the French newspaper *Le Siècle* published a puzzle in which some of the numbers were removed from a magic square and readers had to insert the correct ones. *La France* came very close to inventing sudoku by using magic squares containing only the digits 1–9. In the solutions, 3×3 blocks also contained the nine digits, but this was not made explicit.

Sudoku in its modern form was probably introduced by Howard Garns and published anonymously in 1979 by Dell Magazines under the name 'number place'. In 1986 the Japanese company Nikoli published puzzles of this kind in Japan under the not terribly eye-catching name *sūji wa dokushin ni kagiru* ('the digits are limited to one occurrence'). The name was then shortened to *sū doku*. *The Times* began publishing sudoku puzzles in the UK in 2004, after being contacted by Wayne Gould, who had devised a computer program to produce solutions rapidly. In 2005 it became a worldwide craze.

• •

Hexakosioihexekontahexaphobia

This is the fear of the number 666.

In 1989, when President Ronald Reagan and his wife Nancy moved house, they changed the address from 666 St. Cloud Road to 668 St. Cloud Road. This may not have been a genuine case of hexakosioihexekontahexaphobia, however, because they might not have feared the number *as such*—they were just taking steps to avoid obvious accusations and potential embarrassment.

On the other hand... When Donald Regan, Chief of Staff to President Reagan, published his memoirs in the 1988 *For the Record: From Wall Street to Washington*, he wrote that Nancy Reagan had regularly taken advice from astrologers Jeane Dixon and, later, Joan Quigley. "Virtually every major move and decision the Reagans made during my time as White House Chief of Staff was cleared in advance with a woman in San Francisco who drew up horoscopes to make certain that the planets were in a favorable alignment."

The number 666 has occult significance because it is stated to be the Number of the Beast in the Book of Revelation 13:17–18 (King James translation of the Bible): "And that no man might buy or sell, save he that had the mark, or the name of the beast, or the number of his name. Here is wisdom. Let him that hath understanding count the number of the beast: for it is the number of a man; and his number is Six hundred threescore and six."

It is generally assumed that the reference is to the

numerological system known as *gematria* in Hebrew and *isopsephy* in Greek, in which letters of the alphabet are associated with numbers. Several systems are possible: number the letters of the alphabet consecutively, or number them 1–9, 10–90, and 100-900 or wherever the process stops (which is the ancient Greek number notation). Then the sum of the numbers associated to the letters in a person's name is the numerical value of that name.

Innumerable attempts have been made to deduce who the Beast was. They include the Antichrist (written in Latin in the accusative case as *Antichristum*), the Roman Catholic Church (identified with one of the Pope's titles: *Vicarius Filii Dei*), and Ellen Gould White, founder of the Seventh Day Adventists. How come? Well, counting only the Latin numerals in her name, you get

E	L	L	E	N	G	O	V	L	D	VV	H	I	T	E
	50	50					5		50	500	5+5		1	

adding to 666. If you think the Beast was Adolf Hitler, you can 'prove' it by starting the numbering at $A = 100$:

$$
\begin{aligned}
H &= 107 \\
I &= 108 \\
T &= 119 \\
L &= 111 \\
E &= 104 \\
R &= \underline{117} \\
  &\ 666
\end{aligned}
$$

Basically, you choose your hate-figure, based on your own political or religious views. Then you twist the numbering, and if necessary the name, to fit.

However, all of these deductions may be based on a misapprehension—other than the belief that any of it actually *matters*—because it has now become apparent that 666 may be an error. Around 200 AD the priest Irenaeus knew that several early manuscripts stated a different number, but he attributed this to mistakes by the scribes, asserting that 666 was found in "all the most approved and ancient copies". But in 2005 scholars at Oxford University used computer imaging techniques to read previously illegible portions of the earliest known version of Revelation, number 115 of the papyri found at the ancient site of Oxyrhynchus. This document, which dates to about AD 300, is thought to be the most definitive version of the text. It gives the Number of the Beast as 616.

● ●

Once, Twice, Thrice

The square array

    1 9 2
    3 8 4
    5 7 6

uses each of the nine digits 1–9. The second row 384 is twice the first row 192, and the third row 576 is three times the first row.

There are three other ways to do this. Can you find them?

See page 265 for the answer.

● ●

Conservation of Luck

"A friend of mine won seven million on the Lotto," said the chap next to me in the gym. "That's the end of *my* chances. You can't win if you know someone who has."

There are as many urban myths about the UK National Lottery as there are legs on a millipede, but I'd not come across this one before. It set me wondering: why do people so readily believe this kind of thing?

Think about it. In order for my friend's belief to be true, the Lotto machine has to somehow be influenced by his network of friends and acquaintances. It has to *know* whether any of them has won before, and then take steps to avoid his particular choice of numbers, which means that it also has to know what he has chosen. In fact, all eleven Lotto machines used in the UK lottery must know this, because the one used each week is itself chosen at random.

Since a Lotto machine is an inanimate mechanical device, this doesn't make a great deal of sense.

Each week, the chance of any particular set of six numbers winning the jackpot is 1 in 13,983,816. That's because there are that many possible combinations of numbers, and each is equally likely to occur. If not, the machine would be biased, and it is designed to avoid that. Your chances of winning depend only on what you've chosen this week, not on what someone you know once did. The likely amount you will win does depend on other people, however: if you hit the jackpot and others chose the same number, you all have to share the prize. But that's not what my friend was worried about.

The reasons why some of us believe this kind of myth lie in human psychology rather than probability theory. One possible reason is an unconscious belief in magic, here manifesting itself as luck. If you think that luck is a real *thing* that people possess, and it improves their chances, *and* if you think that there is only a certain amount of luck to go round, then perhaps your fortunate friend has used up all the luck in your neighbourhood. Which in this instance seems to be your social network. OMG! Can you tweet your luck away? Put your luck on Facebook for your so-called friends to steal? It's a nightmare!

Or maybe the underlying idea is like the person who takes a bomb on board an aircraft whenever they travel, on the grounds that the chance of two bombs on the same plane is infinitesimal. (The fallacy is that you chose to bring it on board. This has no effect on the chance of someone else having done the same, unknown to you.)

It's true that most Lottery winners do not have friends who are also winners. So it's easy to deduce that if you want to be a winner, you should avoid having such friends. Actually, most

Lottery winners lack winning friends for the same reason that most losers lack winning friends: there are very few winners but a vast number of losers.

Agreed, you have to be in it to win it. An acquaintance of mine won half a million pounds, and would not have been pleased if I'd advised her not to bother, and then her usual numbers came up.

Knowing that the odds are firmly against me, and not finding the alleged thrill of gambling worth the virtual certainty of throwing my hard-earned money down the drain, I never bet on the Lottery. But I have, over the years, been inadvertently betting on a lottery of my own: writing a bestseller. I haven't won the jackpot, but I've definitely come out ahead. A few years ago the author J.K. Rowling (you know what she's written) became Britain's first self-made female billionaire. That's *five hundred times* the size of a typical Lotto jackpot. And there are a lot fewer than 14 million writers in Britain.

Forget the Lottery. Write a book.

- -

The Case of the Face-Down Aces

From the Memoirs of Dr Watsup

My detective friend suddenly stopped firing his revolver at the chimney breast, having inscribed dotted versions of the letters VIGTO in the plaster. "What *is* it, Watsup?" said he in an irritated tone.

I surfaced from my reverie. "I'm sorry, Soames. Was I disturbing you?"

"I could see you *thinking*, Watsup. The way you purse your lips, and tug at your ear when you believe no one is watching. It is most distracting. One bullet went awry, and now the C looks like a G."

"I was thinking about that new stage magic chap," I said. "Um—"

"The Great Whodunni."

"That's the cove, yes. Clever blighter. Went to see his show last week. Did the most *amazing* card trick, been wondering about it ever since. First he took a pack of cards and dealt the top sixteen face down in four rows of four. Then he turned four of the cards face up. He called for a volunteer from the audience, so of course I put my hand up, but for some reason he chose an attractive young lady instead. Name of Helena . . . well, anyway he told her to repeatedly 'fold up' the square of cards, like folding a sheet of stamps along the perforations, until it ended up as a single pile of sixteen cards."

"She was an accomplice," Soames muttered. "It's elementary."

"I don't think so, Soames. That wouldn't have helped. The audience decided where the folds occurred. For instance, the first fold could be along any of the three horizontal lines between the cards, or the three vertical lines—but the audience called out which."

"The audience was an accomplice, then."

I could tell he was getting into one of his moods. "I chose one of the folds myself, Soames."

The great man nodded distractedly. "Then perhaps the trick was genuine. In which case—ah, yes, the Conundrum of the Concealed Cupcakes springs to mind . . . Tell, me, Watsup: when the cards had been folded into a single pile, was Helena instructed to spread them out on the table? Without turning any over?"

"Yes."

"And did it miraculously transpire that either twelve of the cards were face down and four face up, or four were face down and twelve face up?"

"Yes. The former. And the face-up cards were—"

"The four aces. What else? The whole thing is utterly transparent."

"But it could have been the other way round, Soames," I protested.

"In which case Helena would have been told to take the four face-down cards, and turn them over to reveal . . . "

"Ah. The four aces. I see. But even so, it's an amazing piece of magic. Think of all the different places the aces could be, and all of the ways the audience could have chosen to fold the cards!"

"An amazing piece of chicanery, Watsup."

I allowed my astonishment to show. "You mean—he fixed the audience's votes? Clever psychological trickery?"

"No, Watsup: he fixed the cards. Get me that pack Mrs Soapsuds keeps under the hatstand downstairs for her bridge evening, and I will demonstrate." I hastened to carry out his instructions.

When I reappeared, puffing slightly from my exertions, for I was out of condition, Soames took the cards. He sorted out the four aces, and slid them back into the pack, apparently at random. After dealing out four rows of four cards each, he turned four of the cards over, like this:

Whodunni's initial arrangement

Then he instructed me to fold the pack into a pile, following the same instructions that Whodunni had given to Helena. This done, I spread the cards out, and lo and behold, four were the opposite way up to the other twelve. And those four were...the aces!

"Soames," I cried, "that is truly the most amazing card trick I have ever seen! I do now realise that you must have chosen where to place the four aces, but even so, the number of different ways I could have chosen to fold the cards is huge!"

Soames reloaded his gun. "My dear Watsup, how many times have I told you not to leap to unwarranted conclusions."

"But there really are thousands of ways, Soames!"

The detective gave a curt nod. "That wasn't the conclusion I had in mind, Watsup. Do you really think the choice of folds makes the slightest difference?"

I struck my brow with the palm of my hand. "You mean...it doesn't?" But in reply, Soames merely resumed his attack on the chimney breast.

How does Whodunni's trick work? See page 265 for the answer.

Confused Parents

One of the strangest names for a mathematician is that of Hermann Cäsar Hannibal Schubert (1848-1911) who pioneered enumerative geometry, which counts how many lines or curves defined by algebraic equations satisfy particular conditions. Presumably his parents expected great things of their son, but couldn't work out whose side they were on.

Hermann Cäsar Hannibal Schubert

Jigsaw Paradox

The two triangles appear to have the same area, namely $13 \times 5/2$ = 32.5. But one has a hole in its edge, so the diagram proves that 31.5 = 32.5. What (if anything) is wrong?

See page 267 for the answer.

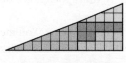 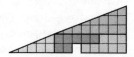

Jigsaw paradox

• •

The Catflap of Fear 🔍

From the Memoirs of Dr Watsup

Hooves skated on the muddy road. The cab screeched round a corner, narrowly missing a barrow loaded with potatoes. The cabbie wiped his brow with a dirty rag.

"Cor, guv! For a moment there I thought we'd 'ad our chips!"*

"Drive on, man! There's a guinea in it for you if you proceed with all undue haste!"

We arrived at our destination. I leaped from the cab, flinging coins at the driver, and rushed past a startled Mrs Soapsuds, up the stairs, and into Soames's lodgings. Without knocking.

"Soames! It's terrible!" I panted. "My—"

"Your cats have been stolen."

"Catnipped, Soames!"

"Surely you mean 'catnapped'?"

---

*   This is not an anachronism: Joseph Malin opened the first fish and chip shop in London in 1860. The delicacy was introduced to Britain in the sixteenth century by Jewish refugees of Spanish and Portuguese origin, under the name *pescato frito*: deep-fried fish. The chips were added later.

"No, they were lured away using a bundle of catnip tied to a piece of string."

"How do you know that?"

"The catnipper left it behind."

Soames gave me a sharp stare. "Unusual. Not like him. Not like him at all."

"Him?"

"Yes. He's back."

I went to the window. "So he is. But this is hardly the time for roasted chestnuts, Soames."

"Watsup, are you out of your mind?"

"The old man who runs the roasted chestnut stall across the street," I explained. "He wasn't there yesterday, but today he is. I assume that is to whom you are referring."

"You *assume*," Soames said scathingly. "Do not assume, Watsup. Examine the evidence and *deduce*." I realised that he was not merely speaking in generalities. There must be something specific that he wished me to deduce.

I pride myself in being unusually sensitive to Soames's moods, and after some reflection I recalled that a few days ago I had come upon him assembling a small arsenal of pistols, rifles, and hand grenades. Now it struck me that perhaps all was not well.

I put this hypothesis to him, and he nodded. "It is as though a ghost from the past has risen from the grave and is sucking the life out of the assembled multitudes of humanity," he said.

"Is it?" I asked. "*What* is, Soames?"

"A foul and dangerous fiend, the Wellington of crime."

"Do you not mean 'Napoleon'? It would seem more apt. The Duke was entirely—"

"He wears rubber boots," Soames explained. "With an extremely common tread pattern, to disguise his footprints. He wears gloves, to leave no fingerprints. He is a master of disguise. He comes and goes without impediment through locked doors. He has the ear of every politician, the eyes of all their wives, and long before that fated day when our paths first crossed he had a

finger in every illicit pie in England. But with superhuman effort
I tracked him down, secured convincing evidence, and broke up
his network of criminal gangs. He fled the country, and I
foolishly thought that was the end of him. But now, I find that
he was merely lying low. He has returned, and he has resumed
his nefarious activities. And now it has become personal."

"Of whom do you speak?"

"Why, Mogiarty! Professor Jim Mogiarty, a brilliant but
flawed mathematician who turned to the Dark Side. He began as
a mere cat burglar, before turning his evil attentions to more
profitable commodities. Not only will he steal anything not
nailed down: he will also steal nails, hammer, and floorboards.
He has dogged my career since—"

"Soames: how can a cat burglar dog anything?"

"Like I said, he is a master of disguise, Watsup. Do listen."

"And how has he manifested himself?"

"Extortion, theft, murder, and kidnap. And now: catnip.
Mogiarty is reverting to his old ways." His expression went grim
with determination. "Never fear, Watsup. We will rescue your
pets—" I glared at him— "your furry feline companions. You
have my word."

I finally thought to ask a vital question. "Soames? How did
you know my cats were missing?" Silently he showed me a torn
envelope. Inside was a scrap of paper and a soggy catnip mouse.

"That's Dysplasia's mouse!" I stifled a manly sob. "What's the
piece of paper about?"

He showed it to me. It read:

CSNSGISTCSTEEVTAOOHAGIAIEITNRETET

"It is a bit of a jumble, Soames, but I see the words STEEV and
HAGIA. Er ... do you know anyone called Stephen in
Constantinople?"

"No, Watsup! It is a code. I have deciphered it."

"How?"

"I observed that there are 33 letters. What does that suggest
to you, Watsup?"

"Er—there wasn't much room on the paper."

"Watsup: 33 is equal to 3×11, a product of two primes. I instantly thought of Mogiarty's mathematical past. And it occurred to me to rearrange the letters in a 3×11 rectangle. Like so."

```
C  S  N  S  G  I  S  T  C  S  T
E  E  V  T  A  O  O  H  A  G  I
A  I  E  I  T  N  R  E  T  E  T
```

He beamed with pride; I could not understand why. It was still gibberish.

"Read down the columns in order, Watsup!"

"CEASEINVESTIGATIONSORTHECATSGETIT. Oh dear!" I was trembling now from head to toe. "Why is Mogiarty doing such a terrible thing to innocent creatures?"

"He is sending us a message."

"That much is clear."

"No, I meant it metaphorically."

"Ah. Has he demanded a ransom?"

"No. I believe it to be a test. I suspect that this crime is merely a trial run for more outrageous ones. He is toying with us like a cat with a mouse."

I stifled another sob. "What can we do?"

"The game is afoot, and we must get ahead of the game, so as not to be taken aback. Already my trusted informants have located your cats in a perfectly ordinary looking house—ironically, in Barking. Actually it is equipped with concealed mantraps, steel doors, bulletproof windows, and alarms of several kinds. There is no possibility of us mounting a clandestine break-in."

I put my service revolver back in my pocket. "A pity."

"However, Mogiarty has made a mistake. There is a boarded-up catflap. We may be able to restore its function and entice your cats through it."

"Yes!" I cried. "I have it! We can tempt them with their favourite treats. Aneurysm likes artichokes, Borborygmus is crazy about banana bread, Cirrhosis can never resist a cream bun, and Dysplasia's downfall is dumplings!"

"Dumplings," said he. "Never mind. A little brainwork, some crucial information, and you see? We progress. We can employ these items to entice the cats to come out through the catflap."

"I have substantial stocks of the necessary comestibles at home," I told him. "I will get them."

"That will indeed be useful, Watsup, when the time comes. But there is a problem. We must present the delicacies in the correct order, because the cats must not be permitted to fight."

"Of course. It could cause injury."

"No, because Mogiarty has filled the basement with high explosive, and rigged it to detonate if the animals fight."

"*What*! Why?"

"Because he has reason to believe that any attempt to rescue the creatures will precipitate a feline altercation. He is using the animals themselves as a warning system. Typically, he ignores the vile consequences of his bloody machinations. As I said, he is sending us a message: that he will stop at *nothing*."

"I see."

"You see, Watsup, but you do not *observe*. Observation begins with enquiries, which provide a context for deduction. I now enquire. In which circumstances do your cats fight? Be precise, for success or failure depends on it."

"Only when they are indoors," I replied, after due reflection.

"Then the house may go up at any moment!"

"No: my cats are entirely peaceable provided certain combinations can be prevented." I wrote down a list of conditions:

- If Cirrhosis and Aneurysm are both indoors, they fight unless Dysplasia is present.
- If Dysplasia and Borborygmus are both indoors, they fight unless Aneurysm is present.
- If Aneurysm and Dysplasia are both indoors, they fight unless Borborygmus or Cirrhosis (or both) are present.

- If Cirrhosis and Dysplasia are both indoors, they fight unless Borborygmus or Aneurysm (or both) are present.
- If Aneurysm or Borborygmus are indoors alone, they won't go out at all.

How can Soames and Watsup entice the cats out without causing an explosion? Only one cat at a time can use the catflap. Ignore trivial moves where a cat goes out and is sent straight back in. If necessary, a cat can be shoved back through the catflap as part of the process.

See page 267 for the answer.

• •

Pancake Numbers

Here is a genuine mathematical mystery—a simple problem whose answer is currently as elusive as the criminal mastermind Mogiarty.

You are given a stack of circular pancakes, all of different sizes. Your task is to rearrange them in order, from the largest at the bottom to the smallest at the top. The only change you are allowed to make is to insert a spatula underneath some pancake in the stack, and use it to pick up the pancakes above it and flip the entire pile over. You can repeat this operation as many times as you like, choosing where to place the spatula as you wish.

Here's an example with four pancakes. It takes three flips to get them in order.

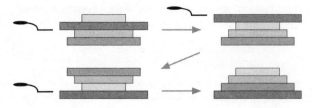

Flipping a stack

Here are some questions for you.

1   Can *every* stack of four pancakes be put in order using at most three flips?

2   If not, what is the smallest number of flips that will put any stack of four pancakes in order?

3   Define the nth pancake number P_n to be the smallest number of flips that will put any stack of n pancakes in order. Prove that P_n is always finite. That is, every stack can be put into the right order using a finite number of flips.

4   Find P_n for $n = 1, 2, 3, 4, 5$. I've stopped at $n = 5$ because there are already 120 different stacks to consider, and to be honest that's an awful lot of work.

See page 269 for the answers, and what else is known.

• •

The Soup Plate Trick

Continuing the cookery theme, there is a curious trick that you can perform using a soup plate, or similar object. Start by balancing the plate on your hand, in the manner of a waiter serving dinner. Now explain that you will accomplish the amazing feat of twisting your arm through a complete turn while keeping the plate horizontal throughout.

To do this, first rotate your arm inwards, tucking the plate under your armpit. Continue to move the plate in a circle, but now raise your arm above your head. Everything returns naturally to the starting position, and the plate doesn't fall off even though you are not gripping it.

You can find videos of the (soup) plate trick on the Internet, for example at

http://www.youtube.com/watch?v=Rzt_byhgujg

where it is referred to as the Balinese cup trick, in reference to Balinese dance using a cup full of liquid instead of a plate. A similar Philippine dance using wine glasses (two per dancer, one in each hand) can be seen on YouTube at

http://www.youtube.com/watch?v=mOO_IQznZCQ

This may seem a fairly trivial manoeuvre, but it has deep mathematical connections. In particular, it helps particle physicists understand one of the curious features of the quantum property known as *spin*. Quantum particles don't *really* spin, like a ball being twirled on a juggler's finger, but there is a number, called spin, which in a certain sense has a similar effect. Spins can be positive or negative, analogous to clockwise and anticlockwise. Some particles have whole number spins: these are called *bosons* (remember the discovery of the Higgs boson?). Others, more bizarrely, have half-integer spins like $\frac{1}{2}$ or $\frac{3}{2}$. These are called *fermions*.

The halves arise because of a very strange phenomenon. If you take a particle with spin 1 (or any integer) and rotate it in space through 360°, it ends up in the same state. But if you take a particle with spin $\frac{1}{2}$ and rotate it in space through 360°, it ends up with spin $-\frac{1}{2}$. You have to rotate it through 720°, *two* full turns, to get the spin back where it began.

The mathematical point here is that there is a 'transformation group' called SU(2), which describes spin and acts by transforming quantum states, and a different group SO(3) that describes rotations in space. These are closely related, but not identical: every rotation in SO(3) corresponds to two distinct transformations in SU(2), one of them being minus the other. This is called a double covering. It's as if SU(2) wraps round SO(3), but it goes round *twice*. A bit like wrapping a rubber band twice round a broomstick.

Physicists illustrate this idea using the Dirac string trick, named for the great quantum physicist Paul Dirac. The idea takes many forms; a particularly simple one uses a ribbon with one end fixed and the other attached to a rotor, which floats in mid-air. The ribbon is shaped like a question mark. After a rotation of 360° the ribbon has not returned to its original position, but to that position rotated through 180°. A second full rotation of the rotor to 720° does not twist the ribbon, but puts it back where it started. The way the ribbon moves is essentially the same as that of the arm holding the soup plate, except that the plate moves around a bit. An astronaut floating in zero gravity could do the same movements with a fixed plate, while keeping his body pointing the same direction at all times.

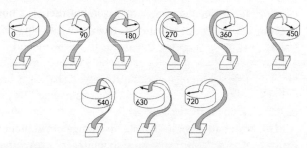

The Dirac string trick using a ribbon. Numbers show the angle in degrees through which the rotor has turned.

A computer-generated movie *Air on Dirac Strings* by George Francis, Lou Kauffman and Daniel Sandin (graphics by Chris Hartman and John Hart) at

http://www.evl.uic.edu/hypercomplex/html/dirac.html

shows the relationship between the Dirac string trick and the Philippine wine dance.

The same idea can be used to connect electrical current to a rotating device such as a wheel. At first sight there's a problem: the wheel has to hover unsupported in mid-air, to allow the ribbon to disentangle. However, in 1975 D.A. Adams designed and patented a device that uses gears to allow the ribbon to rotate completely round the wheel on all sides. It's too complicated to explain here, but see C.L. Stong, The amateur scientist, *Scientific American* (December 1975) 120–125.

Mathematical Haiku

The haiku is a short Japanese verse form, traditionally comprising three separate phrases (lines) that use a total of 17 syllables. The actual Japanese word doesn't correspond exactly to the English concept of a syllable, but that works well enough for haiku in English. The strict traditional pattern uses five syllables in the first and third phrases and seven in the middle one. As an example, here's a haiku by Matsuo Bashō, (1644–1694), in which both the original (omitted) and the translation have that format:

> At the age old pond
> a frog leaps into water
> a deep resonance.

In these decadent modern times, the 5-7-5 pattern is often relaxed, with variations such as 6-5-6 being permitted. In fact, the total of 17 syllables can also be changed. The most important feature is not the precise form, but the emotional content, which requires presenting two distinct but linked images.

The simple format of the haiku has a definite mathematical 'feel', and there are innumerable mathematical haiku. For example:

Daniel Mathews

> Ruler and compass
> Degree of field extension
> Must be power of two.

Jonathan Alperin

> Beautiful theorem
> The basic lemma is false
> Reject the paper.

Jonathan Rosenberg

> Colloquium time.
> Lights out, somebody's snoring.
> Math is such hard work.

Accidental haiku occur when writers unintentionally produce a sentence in haiku format. For example:

> And in the westward
> sky, I saw a curved pale line
> like a vast new moon.

in *The Time Machine* by H.G. Wells. One that Angela Brett noticed (among many others) in the *Princeton Companion to Mathematics* is

> Is every even
> number greater than four the
> sum of two odd primes?

Tim Poston and I dedicated our 1977 *Catastrophe Theory and Its Applications* with a haiku:

> To Christopher Zeeman
> At whose feet we sit
> On whose shoulders we stand.

∗ ∗

The Case of the Cryptic Cartwheel

From the Memoirs of Dr Watsup

Soames was riffling through a stack of newspapers, seeking a crime that would exercise his talents sufficiently to be worth investigating. At that moment, I happened to glance out of the window and saw a familiar figure alighting from a hansom cab. "Why, Soames!" I cried. "It is—"

"Inspector Roulade. He will be here to request our assistance."

There came a knock on the door. I opened it to see Mrs Soapsuds and the Inspector.

"Soames! I've come about—"

"The Downingham kidnapping case. Yes, it does have some features of interest." He passed Roulade the newspaper.

"A sensationalist report, Mr Soames. Ill-informed

speculations about the likely fate of the Earl of Downingham, and the size of the ransom being demanded."

"Predictability of the press," said Soames.

"Yes. Although in this instance it plays into our hands by not revealing certain key facts that might help us to identify—"

"The criminal. Such as the absence of any ransom demand."

"How on Earth—?"

"If there had been a demand, it would by now be public knowledge. It is not. Evidently this is no ordinary kidnapping. We should proceed with all haste to Downingham Hall. Which, if my memory fails me not—and it never does—is on Uppingham Down."

"There is a train from King's Cross to Uppingham in eleven minutes' time," said I, having anticipated his decision and pulled a copy of Bradshaw from the bookcase.

"If we pay the cabbie a guinea, we can just catch it!" cried Soames. "We can discuss the case while we travel."

Upon our arrival at Downingham Hall, the Duke of Southmoreland—who according to the time-honoured rules of the nobility was the father of the Earl of Downingham, who took one of his father's lesser titles—greeted us in person, and quickly conducted us to the scene of the abduction, a muddy paddock outside a barn.

"My son vanished some time during the night," he stated, visibly shaken.

Soames produced his magnifying glass and crawled around in the mud for several minutes. From time to time he muttered to himself. He took out a tape measure and made some measurements in a corner of the barn. Then he rose to his feet.

"I have almost all of the evidence I need," he said. "We must return to London to find the last missing piece." Leaving a baffled Duke standing in his own doorway, next to an equally baffled Inspector, we did just that.

"But, Soames—" I began when we had boarded the train.

"Did you not notice the impression made by the wheels?" he challenged me.

"Wheels?"

"The police had trampled all over the evidence, as usual, but a few traces remained. Enough for me to determine that the Earl departed in a farmyard cart, one of whose wheels fitted tightly against the end of the barn where it meets a high wall. A trace of mud on the wall tells me that a spot on the rim is 8 inches from the ground and 9 inches from the end of the barn. If we can deduce the diameter of the wheel, the case may be close to its solution."

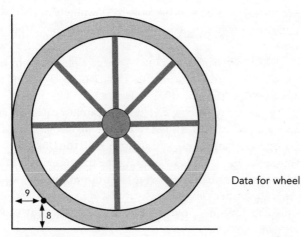

Data for wheel

"*May* be?"

"That depends on the answer. We must also bear in mind that no cartwheel has a diameter of less than 20 inches. Let me see... Ah, yes, it is as I suspected." Upon our arrival at King's Cross station he called for a Baker Street Irreducible—there was always one of the little scamps nearby—and dispatched him to send a telegram to Roulade.

"What does it say?"

"It tells him where the missing Earl can be found."

"But—"

"I know of only one farm in the neighbourhood of Downingham Hall that has a cart with wheels of the exact diameter that I have calculated, which is distinctively large. I am convinced that the Earl left the Hall voluntarily under cover of darkness, using a humble cart to avoid attracting attention. He will be at the place where the cart is habitually kept."

Next morning Mrs Soapsuds brought a telegram from the Inspector: EARL OF D SAFE AND WELL CONGRATULATIONS ROULADE.

"So where did the Earl go?" I asked eagerly.

"That, Watsup, is a secret whose disclosure would destroy the reputations of several of the most revered families in Europe. But I *can* tell you the size of the wheel."

What was the wheel's diameter? See page 271 for the answer.

Two by Two

There are thousands of Noah's ark cartoons. My favourite has a biological theme. The last few pairs of animals—elephants, giraffes, monkeys—are being loaded up the ramp into the ark. Noah is grubbing around on the ground on hands and knees. His wife is leaning over the side of the ark, yelling "Noah! Forget the other amoeba!"

There's a mathematical Noah's ark joke, well worn but perfectly crafted.

As the flood recedes, Noah lets all the animals loose, telling them to go forth and multiply. After a year or so, he decides to check up on them. There are baby elephants, rabbits, goats, crocodiles, giraffes, hippos, and cassowaries everywhere. But then he comes across a lone pair of snakes, looking dejected.

"What's the problem?" asks Noah.

"Can't multiply," says one of the snakes. (Bear in mind that Noah is a sort of Dr Dolittle figure and can talk to the animals.)

A passing chimpanzee overhears the conversation. "Cut down some trees, Noah."

Noah is puzzled but does as the chimp says. A few months

later he visits the snakes, and now there are little snakes everywhere and everyone is happy.

"OK, how did that happen?" Noah asks the snakes.

"We're adders. We can only multiply using logs."

• •

The V-shaped Goose Mystery

Migrating flocks of birds often fly in a V-shaped formation. V-shaped skeins of geese are especially familiar, and they often contain dozens or even hundreds of birds. What makes them adopt this shape?

Researchers have long suggested that this formation saves energy by avoiding birds getting caught up in the turbulent wake of those in front, and recent experimental and theoretical studies have confirmed this general viewpoint. But this theory relies on the birds being able to sense the air currents and adjust their flight accordingly, and until recently it's not been clear that they can do this.

An alternative explanation is that the flock has a leader—the one in front—and everyone else follows the leader. Perhaps the leader is the best navigator, the one who knows where to go. Perhaps it's just whichever bird finds itself at the front.

Birds in V-formation, flying from right to left. Mostly. Where's the bird at top right going? There's always *one*...

Before proceeding to the answer, we need to understand a few basic features of bird flight. In steady flight, a bird flaps its wings in a repetitive cycle, down-beat followed by up-beat. It gains lift from the down-beat as vortices of air spin off from the edges of the wings, and it uses the up-beat to return the wing to its original position, so that the cycle can repeat. The length of the cycle is called the period.

Suppose that two birds are flying using cycles of the same period, which is pretty much what happens in a migrating flock. Although they move in the same way, they need not make the same motions at the same time. For example, when one bird is producing a down-beat, the other may be on an up-beat. The relation between their timing is called the relative phase, and it is the fraction of a cycle between one bird starting its down-beat and the other bird starting its own down-beat.

Thanks to some remarkable detective work by Steven Portugal and his team, we now know that the energy-saving theory is correct, *and* that the birds can indeed sense the invisible air currents well enough to carry it out. The big problem for experimental studies is that the birds you are trying to observe rapidly disappear from view, along with any attached equipment.

Enter the bald ibis.

Once there were so many bald ibises that the ancient Egyptians used a stylised picture as their hieroglyph for *akh*, meaning 'to shine'. Today only a few hundred survive, mainly in Morocco. So a captive breeding programme has been set up at a zoo in Vienna. A lot of effort goes into teaching the birds to follow the correct migration routes. This is done by training them to follow a microlight aircraft, which is flown along parts of the route—but also returns to base, along with the birds.

Portugal realised that it would be possible to make extensive measurements of the positions of the birds, and how they move their wings, from the aircraft. Instead of the birds disappearing over the horizon, they stay close to the equipment. What his team found was amazing and elegant. Each bird positions itself behind and slightly to the side of the one in front, and it adjusts the relative phase of its wing flaps so that it rides on the updraught created by the vortex spun off by the bird in front. The second bird must not only get its wingtip into the right place, which is relatively tiny, but it must also adjust the phase of its flap to exploit the updraught efficiently.

Wingtip placement and phase adjustment.
Grey curves: vortices spun off from wingtips.
Arrows show rotation of vortex.

At first sight these considerations would also permit a zigzag formation, in which each bird flies to one side of the one in front, but not forming a single V shape. (It has a choice of flying to the left or to the right.) However, the first bird to break the V shape (say by flying to the right of the bird ahead rather than on the outside of the V to its left) would be directly behind the bird two places in front of it. The air there would be turbulent, disturbed by the bird directly in front, so it would be much harder to extract lift by placing the wingtip correctly. This problem is avoided by flying on the outside of the V where the air is undisturbed.

It would also be possible for the birds to form a single diagonal line, like one arm of the V. But this would leave room for birds to join the other arm, closer to the leader. However, it is common for one arm of the V to be longer than the other.

Why not fly in a more complex zigzag formation, like this or something similar but more wiggly?

In experiments with ibises, it took quite a while for juvenile birds to learn how to position themselves. In practice a few birds may get it wrong, and the V shape is seldom perfect. Nonetheless, the detailed experiments show conclusively that ibises are capable of sensing the air flow well enough to position themselves in or close to the most energy-efficient location relative to the bird in front.

See page 272 for further information.

• •

Eelish Mnemonics

There are innumerable mnemonics for π (see pages 39–40). Mnemonics for that other famous mathematical constant, the base of natural logarithms

e = 2.7182818284 5904523536 0287471352662497757 . . .

are rarer. Among them are two that give ten figures:

> To disrupt a playroom is commonly a practice of children. It enables a numskull to memorise a quantity of numerals.

There's also a 40-digit self-referential mnemonic devised by Zeev Barel [A mnemonic for e, *Mathematics Magazine* 68 (1995) 253], which you should compare with the decimal expansion above. It uses an exclamation mark '!' in quotes to represent 0, and it goes like this:

We present a mnemonic to memorise a constant so exciting that Euler exclaimed: '!' when first it was found, yes, loudly '!'. My students perhaps will compute e, use power or Taylor series, an easy summation formula, obvious, clear, elegant.

The 'easy summation formula' is

$$e = 1 + \frac{1}{1!} + \frac{1}{2!} + \frac{1}{3!} + \frac{1}{4!} + \frac{1}{5!} + \cdots$$

going on forever. Now ! denotes the factorial

$$n! = n \times (n - 1) \times \cdots \times 3 \times 2 \times 1$$

Since π mnemonics are written in Pilish [page 39], are e mnemonics written in Eelish?

Amazing Squares

There are infinitely many natural numbers that can be expressed as sums of three squares in two different ways: $a^2 + b^2 + c^2 = d^2 + e^2 + f^2$. But more is achievable. An amazing example is

$$123789^2 + 561945^2 + 642864^2 = 242868^2 + 761943^2 + 323787^2$$

The relationship is preserved if we repeatedly delete the leftmost digit:

$$23789^2 + 61945^2 + 42864^2 = 42868^2 + 61943^2 + 23787^2$$
$$3789^2 + 1945^2 + 2864^2 = 2868^2 + 1943^2 + 3787^2$$
$$789^2 + 945^2 + 864^2 = 868^2 + 943^2 + 787^2$$
$$89^2 + 45^2 + 64^2 = 68^2 + 43^2 + 87^2$$
$$9^2 + 5^2 + 4^2 = 8^2 + 3^2 + 7^2$$

It is also preserved if we repeatedly delete the rightmost digit:

$$12378^2 + 56194^2 + 64286^2 = 24286^2 + 76194^2 + 32378^2$$
$$1237^2 + 5619^2 + 6428^2 = 2428^2 + 7619^2 + 3237^2$$
$$123^2 + 561^2 + 642^2 = 242^2 + 761^2 + 323^2$$
$$12^2 + 56^2 + 64^2 = 24^2 + 76^2 + 32^2$$
$$1^2 + 5^2 + 6^2 = 2^2 + 7^2 + 3^2$$

or if we delete digits from both ends:

$$2378^2 + 6194^2 + 4286^2 = 4286^2 + 6194^2 + 2378^2$$
$$37^2 + 19^2 + 28^2 = 28^2 + 19^2 + 37^2$$

This mathematical mystery was sent to me by Moloy De and Nirmalya Chattopadhyay, who also explained the simple but clever idea involved. Can you emulate Hemlock Soames and ferret out the secret?

See page 273 for the answer.

• •

The Thirty-Seven Mystery

From the Memoirs of Dr Watsup

"How curious!" I remarked, thinking aloud.

"Many things are curious, Watsup," said Soames, whom I had thought to have fallen asleep in his chair. "Which curiosity do you have in mind?"

"I took the number 123 and repeated it six times," I explained.

"Getting 123123123123123123," Soames said dismissively.

"Ah, yes, but I haven't finished."

"You multiplied that by 37, no doubt," said the great detective, once again confounding my expectation that I could tell him anything that he did not know already.

"Yes! I did! And what I got was—no, Soames, please do not interrupt—the answer

4555555555555555551

with a great many repetitions of the digit 5."

"And that is curious?"

"Undoubtedly. While one such calculation might be mere coincidence, something similar happens if I use 234 or 345 or 456 instead of 123. Look!" And I showed him my arithmetic:

$$234234234234234234 \times 37 = 8666666666666666658$$
$$345345345345345345 \times 37 = 12777777777777777765$$
$$456456456456456456 \times 37 = 16888888888888888872$$

"Not only that: if I repeat 123 or 234 or 345 or 456 a different number of times, and multiply by 37, I again obtain a great many repetitions of the same digit, except near the ends."

"I am inclined to think," murmured Soames, "that the patterns 123, 234, 345, and so on, are irrelevant. Have you tried other numbers?"

"I tried 124, and that didn't work. Look!"

$$124124124124124124 \times 37 = 4592592592592592588$$

"Digits repeat in blocks of three, but I don't find that surprising, since they do the same in the first number."

"Did you try 486?"

"No—well, since 124 fails, I really don't think... Oh, very well." I returned to my notebook and wrote down the calculation. "How curious!" I said again, having discovered the answer:

$$486486486486486486 \times 37 = 17999999999999999982$$

Inspired, I tried various three-digit numbers at random, writing them down six times in a row and multiplying by 37. Sometimes the result included many repetitions of the same digit, but more often, it did not. I showed Soames my working, and confessed: "I am flummoxed."

"The mystery will no doubt resolve itself," Soames replied, "if you consider the number 111."

I wrote down

$$111111111111111111 \times 37 = 4111111111111111107$$

and stared at it. After twenty minutes had passed, Soames got up and glanced over my shoulder. He shook his head in amusement. "No, no, Watsup! I was not suggesting you try your method on the number 111!"

"Oh. I assumed—"

"How many times have I told you, Watsup: *do not assume anything*! Although the mystery seems to involve the number 37, that is something of a side issue. I was suggesting that you should contemplate how the number 111 relates to 37."

See page 273 for the answer.

• •

Average Speed

Because of heavy traffic, a bus drives from Edinburgh to London, a distance of 400 miles, in 10 hours, a speed of 40 mph. It makes the return journey in 8 hours, a speed of 50 mph. What is its average speed for the whole journey?

The obvious answer is 45 mph, the arithmetic mean of 40 and 50, obtained by adding them and dividing by 2. However, the bus performs the total round trip of 800 miles in 18 hours, an average speed of $800/18 = 44\frac{4}{9}$ mph.

How come?

See page 275 for the answer.

• •

Four Clueless Pseudoku

The clueless puzzle on page 41 was created by Gerard Butters, Frederick Henle, James Henle, and Colleen McGaughey. It is a variant of sudoku which I like to call clueless pseudoku. Here are four more clueless pseudoku mysteries to solve. The rules are:

● Each row and each column must contain each number 1, 2, 3,..., *n* once only, where *n* is the size of the square.

● The numbers in each of the regions outlined in thick black
   lines must have the same total. I've written it above the
   puzzle to save you the bother of working it out. Each solution
   is unique except for the fourth puzzle, where there are two
   solutions, symmetrically related.

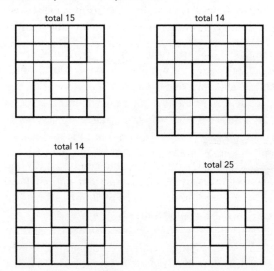

total 15   total 14

total 14   total 25

Four clueless pseudoku mysteries

See page 276 for the answers and further reading.

● ●

Sums of Cubes

Triangular numbers 1, 3, 6, 10, 15, and so on, are defined by
adding consecutive numbers together, starting from 1:

$$1 = 1$$
$$1 + 2 = 3$$
$$1 + 2 + 3 = 6$$
$$1 + 2 + 3 + 4 = 10$$
$$1 + 2 + 3 + 4 + 5 = 15$$

and so on. There is a formula

$$1 + 2 + 3 + \cdots + n = n(n+1)/2$$

and one way to prove it is to write the sum *twice*, like this:

$$1 + 2 + 3 + 4 + 5$$
$$5 + 4 + 3 + 2 + 1$$

and observe that the numbers in vertical columns all add to the same thing, here 6. So twice the sum is $6 \times 5 = 30$, and the sum is 15. If you did this with the numbers from 1 to 100, it would work in much the same way: there would be 100 columns each adding to 101, so the sum of the first 100 numbers must be half of 100×101, which is 5,050. More generally, if we add the first n numbers we get half of $n(n+1)$, and that's the formula.

There's a formula for sums of squares, but it's a bit more complicated:

$$1 + 4 + 9 + \cdots + n^2 = n(n+1)(2n+1)/6$$

But what happens for *cubes* is very striking:

$$1^3 = 1$$
$$1^3 + 2^3 = 9$$
$$1^3 + 2^3 + 3^3 = 36$$
$$1^3 + 2^3 + 3^3 + 4^3 = 100$$
$$1^3 + 2^3 + 3^3 + 4^3 + 5^3 = 225$$

The results are the squares of the corresponding triangular numbers.

Why should sums of cubes give squares? We can find the formula and prove it that way, but there's a very neat pictorial proof that

$$1^3 + 2^3 + 3^3 + \cdots + n^3 = (1 + 2 + 3 + \cdots + n)^2$$

without using any formulas.

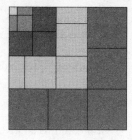

Visualising sums of cubes

The picture shows 1 square of side 1, 2 of side 2 (making a $2 \times 2 \times 2$ cube), 3 of side 3 (a $3 \times 3 \times 3$ cube), and so on. So the total area is the sum of consecutive cubes. Reading along the top edge we find $1 + 2 + 3 + 4 + 5$, the sum of consecutive numbers. But the area of a square is the square of the length of its side. Done!

If you want a formula, we know that $(1 + 2 + 3 + \cdots + n) = n(n+1)/2$, so squaring gives $1^3 + 2^3 + 3^3 + \cdots + n^3 = n^2(n+1)^2/4$.

• •

The Puzzle of the Purloined Papers

From the Memoirs of Dr Watsup

Soames passed me an envelope, and held up the letter that it had enclosed.

"A test of your powers of observation, Watsup. Who do you think sent me this?"

I held it to the light, looked at the postmark and stamp, sniffed, examined the glue where it had been sealed. "The sender is a woman," I said. "Unmarried but not yet condemned to spinsterhood, and actively seeking a husband. She is frightened, but courageous." I paused, and further inspiration struck. "Her finances are strained, but not yet disastrous."

"Very good," said he. "I see you have absorbed some of my methods."

"I do my best," said I.

"Explain what led you to these deductions."

I ordered my thoughts. "The envelope is pink, and it bears

distinct traces of perfume. *Nuits de Plaisir*, if I am not mistaken, for my friend Beatrix often wears the same. It is too daring for a married woman, not daring enough for a young one. That she wears perfume at all implies she is actively seeking male attention. Traces of cosmetics on the flap confirm this. But the glue had been licked only partially, suggesting that her mouth was dry when she sealed the envelope, and a dry mouth is a sign of fear. Since she nevertheless completed the task and posted the letter to you, she is still able to function rationally, under severe stress, a sign of courage.

"Finally, the stamp shows signs of having been steamed off another envelope and been reused—a bent corner, traces of a previous postmark. This indicates a frugal attitude. However, she can afford perfume, so she is not yet on poverty's doorstep."

He nodded thoughtfully, and I mentally preened myself.

"There are a few signs you missed," he said quietly, "which put the matter in a fresh light. The shape and size of the envelope reveal it as government issue, not to be found in any high-street stationer's. I refer you to my monograph on stationery and its characteristic dimensions. The ink used to write the address is a strange shade of dark brown, unavailable commercially but provided in bulk to certain departments in Whitehall."

"Ah! Then her current beau is a civil servant and she borrowed both envelope and ink from him."

"A sensible theory," said he. "Totally incorrect, of course, but eminently sensible and consistent with much of the evidence. However, in fact this letter is from my brother Spycraft."

I was shocked rigid. "You have a brother?" Soames had never talked of his family.

"Oh, have I not mentioned him? How very remiss of me."

"How do you know he wrote the letter?"

"He signed his name to it."

"Oh. But what of the other clues?"

"Spycraft's little joke. But we must make haste, for we are to meet him forthwith at the Diophantus Club. Give an urchin sixpence to call a hansom, and I will tell you more as we make our way there."

As we clattered along Portland Place, he explained that his brother was a retired expert on prime numbers, who did occasional freelance work for Her Majesty's Government. He refused to be drawn on the nature of the work, saying only that it was highly confidential and politically sensitive.

Upon arriving at the Diophantus Club, we were ushered into the Visitor's Lounge, where a gentleman was waiting in a comfortable armchair. My immediate impression was one of languid corpulence, but it concealed a sharpness of mind and alertness of body that belied this initial assessment.

Soames introduced us.

"You often find my own deductive abilities astonishing, Watsup," he said, "but Spycraft puts me to shame."

"There is one area where your abilities exceed my own," his brother contradicted. "Namely, logical conundrums in which the precise conditions are fluid. I find that I have no basis from which to attack the question. Whence my note."

"I take it that you have no objection to Dr Watsup being told all?"

"His service record in Al-Jebraistan is impeccable. He must be sworn to secrecy, but his word will be sufficient."

Soames gave his brother a sharp look. "It's not like you to accept someone's word."

"It will be sufficient when I inform him of the consequences of breaking it."

I duly swore, and we got down to business.

"An important document was accidentally mislaid, and then stolen," Spycraft said. "It is essential to the security of the British Empire that it be recovered without delay. If it gets into the hands of our enemies, careers will be ruined and parts of the Empire may fall. Fortunately, a local constable caught a glimpse of the thief, enough to narrow it down to precisely one of four men."

"Petty thieves?"

"No, all four are gentlemen of high repute. Admiral Arbuthnot, Bishop Burlington, Captain Charlesworth, and Doctor Dashingham."

Soames sat bolt upright. "Mogiarty has a hand in this, then."

Not following his reasoning, I asked him to explain.

"All four are spies, Watsup. Working for Mogiarty."

"Then . . . Spycraft must be engaged in counter-espionage!" I cried.

"Yes." He glanced at his brother. "But you did not hear that from me."

"Have these traitors been questioned?" I asked.

Spycraft handed me a dossier, and I read it aloud for Soames's benefit. "Under interrogation Arbuthnot said 'Burlington did it.' Burlington said 'Arbuthnot is lying.' Charlesworth said 'It was not I.' Dashingham said 'Arbuthnot did it.' That is all."

"Not quite all. We know from another source that exactly one of them was telling the truth."

"You have an informer in Mogiarty's inner circle, Spycraft?"

"We *had* an informer, Hemlock. He was garrotted with his own necktie before he could tell us the actual name. Very sad—it was an Old Etonian tie, totally ruined. However, all is not lost. If we can deduce who was the thief, we can obtain a search warrant and recover the document. All four men are being watched; they will have no opportunity to pass the document to Mogiarty. But our hands are tied; we must stick to the letter of the law. Moreover, if we raid the wrong premises, Mogiarty's lawyers will publicise the mistake and cause irreparable damage."

Which man was the thief? See page 276 for the answer.

• •

Master of All He Surveys

A farmer wanted to enclose as large an area of field as possible, using the shortest possible fence. Perhaps unwisely, he called the local university, who sent an engineer, a physicist, and a mathematician to advise him.

The engineer built a circular fence, saying it was the most efficient shape.

The physicist built a straight line so long that you couldn't see the ends, and told the farmer that to all intents and purposes it went right round the Earth, so he had fenced in half the planet.

The mathematician built a tiny circular fence around himself and said "I declare myself to be on the outside."

* *

Another Number Curiosity

$$1 \times 8 + 1 = 9$$
$$12 \times 8 + 2 = 98$$
$$123 \times 8 + 3 = 987$$
$$1234 \times 8 + 4 = 9876$$
$$12345 \times 8 + 5 = 98765$$

So, all you budding Hemlock Soameses: what comes next and when does the pattern stop?

See page 278 for the answers.

* *

The Opaque Square Problem

Speaking of fences: what is the shortest fence that will block every line of sight across a square field? That is, a fence that meets every straight line that intersects the field. This is the Opaque Square Problem: the name indicates that you can't see through it. The question goes back to Stefan Mazurkiewicz in 1916, who asked it for any shape, not just a square. It remains baffling, but some progress has been made.

Suppose the side of the field is one unit. Then a fence round all four sides would certainly work, and the length would be 4. However, we could cut out one of the sides and still have an opaque fence, reducing the answer to 3. This is the shortest fence formed by a single polygonal line. But if we allow fences using several lines, a shorter possibility quickly springs to mind: the two diagonals of the field, total length $2\sqrt{2} = 2.828$, approximately.

Can we do better still? One general fact is clear: an opaque fence contained entirely within the field must include all four corners of the square. If some corner were not included, there would be a line intersecting the square only at that corner (cutting diagonally through it from outside) and that would miss the fence. Even one such line breaks the conditions of the problem.

Any fence that includes all four corners and joins them together must be opaque, because any line that intersects the square must either pass through a corner or separate two of them. Then any connecting fence must cross that line. Is the pair of diagonals the shortest such fence? No. The shortest fence that connects all four corners, called a Steiner tree, has length $1 + \sqrt{3}$ = 2.732, approximately. The lines meet at angles of 120°.

It turns out that even this fence is not the shortest opaque one. There is a disconnected fence, in which one part blocks lines of sight through the gap, of length $\sqrt{2} + \sqrt{3/2} = 2.639$. It is widely believed, but not yet proved, that this is the shortest opaque fence. Bernd Kawohl has proved that this is the shortest fence that has exactly two connected pieces. One is the Steiner tree linking three corners, three lines that meet at 120°; the other is the shortest straight line between the centre and the fourth corner.

Opaque fences for a square. *From left to right*: lengths are 4, 3, 2.828, 2.732, and 2.639.

We don't even know for sure that there *is* a shortest opaque fence. Or that if there is one, it must be entirely within the square. Vance Faber and Jan Mycielski have proved that for any given finite number of pieces, at least one shortest opaque fence exists. (For all we know, there might be several.) The technical problem here, which is currently unsolved, is the possibility that the more pieces you permit, the shorter the fence can be. It would then be possible to find a series of fences with ever-shorter lengths, but no fence that is shorter than all of these. Alternatively, a fence composed of infinitely many disconnected pieces might be the shortest.

● ●

Opaque Polygons and Circles

A standard mathematician's trick, when you can't solve a problem, is to generalise it: consider a range of similar but more complicated problems. This might seem a stupid idea: how can making the question *harder* help you to solve it? But the more examples you have to think about, the better your chances of spotting some interesting common feature that cracks the problem. It doesn't always work, and so far it hasn't done here, but occasionally it does.

One way to generalise the Opaque Square Problem is to change the shape. Replace the square by a rectangle, or a polygon with more sides, a circle, an ellipse—the possibilities are endless.

Mathematicians have mainly concentrated on two generalisations: regular polygons and circles. The shortest known opaque fence for the equilateral triangle is a Steiner tree joining each corner to the centre by a straight line. There is a general construction that gives the shortest known opaque fences for regular polygons with an odd number of sides, and a similar but different one for an even number of sides.

Shortest known opaque fences for regular polygons. *From left to right*: equilateral triangle; odd-sided regular polygon; even-sided regular polygon.

What about an opaque circle? If the fence has to stay within the shape, the obvious answer is the perimeter of the circle. If it's a unit circle, this has length $2\pi = 6.282$. If part of the perimeter is missing, you need extra bits of fence inside the circle to block paths that cross this missing segment, and it gets complicated. Intuitively, a circle can be thought of as a regular polygon with infinitely many infinitely short sides. Based on this idea, Kawohl has proved that a construction like that for regular polygons, but using an infinite number of pieces, gives an opaque fence of total length $\pi + 2 = 5.141$, which is smaller than 2π. But there is a shorter U-shaped opaque fence if some of it can lie outside the circle. This also has length $\pi + 2$, and is conjectured to be the shortest possible, and it has been proved to be so for fences formed by a single curve with no branch points.

Opaque fences for a circle. *Left*: Obvious but not the best. *Right*: A shorter opaque fence that goes outside the circle.

The problem has also been extended to three dimensions: now the fence has to be a surface, or something more complicated. The best known opaque fence for a cube is formed from several curved pieces.

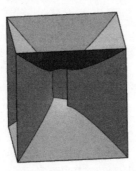

The best known opaque fence for a cube.

πr^2?

No, pie are round. Chocolate are squared.

• •

The Sign of One 🔍

From the Memoirs of Dr Watsup

"Soames! Here's a pretty puzzle. It might interest you."

Hemlock Soames put down his clarinet, upon which he had been playing a Bolivian funeral dirge. "I doubt it, Watsup." He had been in this melancholy mood for some weeks, and I was determined to kick him out of it.

"The problem is to express the integers 1, 2, 3, and so on, using at most—"

"Four 4's," said Soames. "I know it well, Watsup.*"

I decided not to let his lack of enthusiasm daunt me. "With basic arithmetical symbols it is possible to reach 22. Square roots increase this limit to 30. Factorials raise it to 112; powers to 156—"

"And subfactorials to 877," Soames finished. "It is an old puzzle, and one that has been wrung dry."

"What's a subfactorial, Soames?" I asked, but he had already buried his nose in yesterday's *Daily Wail*.

After a moment, he reappeared. "Mind you, there are many possible variations. The use of 4 allows considerable freedom, and several useful numbers can be created using just one 4. Such as $\sqrt{4} = 2$ and $4! = 24$."

"What does the exclamation mark mean," I asked.

"Factorial. For instance, $4! = 4 \times 3 \times 2 \times 1$, and so on. Which, as I said, is 24."

"Oh."

"These extra numbers come free of charge, and so make the puzzle easier. But I wonder..." His voice trailed off.

* W.W. Rouse Ball, *Mathematical Recreations and Essays* (11th edition), Macmillan, London 1939.

"Wonder what, Soames?"

"I wonder how far one can get using four *ones*."

Silently I rejoiced, for his interest was clearly piqued. "Yes, I see," I said. "Now $\sqrt{1} = 1$ and $1! = 1$, so no new numbers arise 'for free'. Which makes the problem harder, and perhaps more worthy of our attention."

He grunted, and I hastened to press home my slight advantage. The best way to interest Soames in a problem was to attempt it, and fail.

"I see that

$$1 = 1 \times 1 \times 1 \times 1$$

and

$$2 = (1 + 1) \times 1 \times 1$$
$$3 = (1 + 1 + 1) \times 1$$
$$4 = 1 + 1 + 1 + 1$$

But an expression for 5 escapes me."

Soames raised one eyebrow. "You might consider

$$5 = (1/.1)/(1 + 1)$$

where the dot is the decimal point."

"Oh, that's rather clever!" I cried, but Soames merely snorted. "So what about 6?" I continued. "I can see how to do it using factorials:

$$6 = (1 + 1 + 1)! \times 1$$

I really only need three 1's, but any spares can always be got rid of by multiplying by 1."

"Elementary," he muttered. "Have you considered

$$6 = \sqrt{1/.\dot{1}} + \sqrt{1/.\dot{1}}$$

Watsup? Or, for that matter,

$$6 = \left(\sqrt{1/.\dot{1}} \right)!$$

if you insist on employing factorials. You may multiply by 1×1 or $1/1$, or add $1 - 1$, to use all four 1's, of course."

I stared at the formula. "I recognise the decimal point, Soames, but what is the extra $\dot{1}$?"

"Recurring," Soames replied wearily. "Nought point one recurring is equal to $0.111111\ldots$ going on forever. Of course the initial zero may be omitted. The infinite recurring decimal equals $1/9$ *exactly*. Dividing that into 1 gives 9, whose square root is 3—"

"And then $3 + 3 = 6$," I yelled excitedly. "And, of course,

$$7 = (1 + 1 + 1)! + 1$$

is easy without square roots. But 8 is a kettle of worms of a different colour—"

"Do pay attention," said Soames.

$$8 = 1/.\dot{1} - 1 \times 1$$
$$9 = 1/.\dot{1} + 1 - 1$$

"Ah! Yes! And then

$$10 = 1/.\dot{1} + 1 \times 1$$
$$11 = 1/.\dot{1} + 1 + 1$$

and..."

"You are using up 1's prodigiously," said Soames. "It is best to save them for later." And he wrote down

$$10 = 1/.1$$
$$11 = 11$$

adding, "Note the absence of the 'recurring' symbol, Watsup. This time it is just the ordinary decimal .1. Oh, and you must multiply both by 1×1 to use up the spare 1's, or do so in one of the other ways I remarked upon. But later, you can omit those two 1's and put them to good purpose."

"Yes! You mean, like

$$12 = 11 + 1 \times 1$$
$$13 = 11 + 1 + 1$$
$$14 = 11 + \sqrt{1/.\dot{1}}$$

and so on?"

A flicker of a smile crossed Soames's countenance. "You have it, Watsup!"

"But what about 15?" I asked.

"Trivial," he sighed, and wrote down

$$15 = 1/.\dot{1} + \left(\sqrt{1/.\dot{1}}\right)!$$

To which I triumphantly appended

$$16 = 1/.1 + \left(\sqrt{1/.\dot{1}}\right)!$$
$$17 = 11 + \left(\sqrt{1/.\dot{1}}\right)!$$
$$18 = 1/.\dot{1} + 1/.\dot{1}$$
$$19 = 1/.1 + 1/.\dot{1}$$
$$20 = 1/.1 + 1/.1$$
$$21 = 1/.1 + 11$$
$$22 = 11 + 11$$

and Soames nodded approvingly. "Now it starts to become interesting," he remarked. "What of 23, I wonder?"

"I've got it, Soames!" I cried:

$$23 = (\sqrt{1/.\dot{1}} + 1)! - 1$$
$$24 = (\sqrt{1/.\dot{1}} + 1)! \times 1$$
$$25 = (\sqrt{1/.\dot{1}} + 1)! + 1$$

Bearing in mind," I clarified, "that $4! = 24$, as you so sagely remarked. This is fun, Soames! Though for the life of me I can't manage to express 26."

"Well..." he began, and paused.

"Stumped, you, has it?"

"Not in the least. I was merely wondering whether it is necessary to introduce a new symbol. It will certainly make life easier. Watsup, are you aware of the floor and ceiling functions?"

My gaze inadvertently went to my feet, and then up above my head, but no inspiration struck.

"I see you do not," said Soames. *How does he know what I'm thinking?* I thought. *It's—*

"Uncanny...yes, is it not? I read you like a book, Watsup. Possibly *Mother Goose*. Now, these functions are

$$\lfloor x \rfloor = \text{the greatest integer less than or equal to } x \text{ (floor)}$$
$$\lceil x \rceil = \text{the smallest integer greater than or equal to } x$$
$$\text{(ceiling)}$$

and you will find them indispensable in all puzzles of this sort."

"Excellent, Soames. Though I admit, I fail to see..."

"The idea, Watsup, is that by their means we can express useful small numbers using only *two* 1's. For example,

$$3 = \left\lfloor \sqrt{1/.1} \right\rfloor$$

is another way to represent 3 with two 1's, and

$$4 = \left\lceil \sqrt{1/.1} \right\rceil$$

is new." Seeing my bewilderment, he added: "You appreciate that $\sqrt{1/.1} = \sqrt{10} = 3.162$, whose floor is 3 and ceiling is 4."

"Yes..." I said doubtfully.

"Then we advance, because

$$26 = \left\lceil \sqrt{1/.1} \right\rceil! + 1 + 1$$
$$27 = \left\lceil \sqrt{1/.1} \right\rceil! + \left\lfloor \sqrt{1/.1} \right\rfloor$$
$$28 = \left\lceil \sqrt{1/.1} \right\rceil! + \left\lceil \sqrt{1/.1} \right\rceil$$

Not to mention various alternatives."

Thousands of incoherent thoughts surged through my brain. One stood out. "Why, Soames, I've just realised that

$$5 = \left\lceil \sqrt{\left\lceil \sqrt{1/.1} \right\rceil !} \right\rceil$$

because $\sqrt{24} = 4.89$, whose ceiling is 5. So I can now make 29 and 30!" By which I meant 30, not factorial 30, you understand. Punctuation is such a pain.

Watsup and Soames investigated the puzzle much further, and we will see what they discovered later. But before continuing the story, you might like to see how far you can get on your own. Starting with 31.

The Sign of One continues on page 105.

. . . .

Progress on Prime Gaps

Recall that a whole number is *composite* if it can be obtained by multiplying two smaller whole numbers together, and *prime* if it can't be obtained by multiplying two smaller whole numbers together and it is greater than 1. The number 1 is exceptional: a few centuries ago it was considered to be prime, but this convention stops prime factors being unique. For example, $6 = 2 \times 3 = 1 \times 2 \times 3 = 1 \times 1 \times 2 \times 3$, and so on. Nowadays, for this reason and others, 1 is considered to be special. It is neither prime nor composite, but a *unit*: a whole number x such that $1/x$ is also a whole number. Indeed, it is the only positive unit.

The first few primes are

2 3 5 7 11 13 17 19 23 29 31 37

There are infinitely many of them, scattered in an irregular manner throughout the whole numbers. The prime numbers have long been a huge source of inspiration, and many of their mysteries have been solved over the years. But many others remain as opaque as they have ever been.

In 2013 number theorists made sudden and unexpected progress on two of the great mysteries about prime numbers. The first concerned the gaps between successive primes, and I'll describe that now. The second follows immediately after.

All primes other than 2 are odd (since all even numbers are multiples of 2), so it is not possible for two consecutive numbers except (2, 3) both to be prime. However, it is possible for two numbers that differ by 2 to be prime: examples are (3, 5), (5, 7), (11, 13), (17, 19), and it is easy to find more. Such pairs are called *twin primes*.

It has long been conjectured that there are infinitely many twin prime pairs, but this has not yet been proved. Until recently progress on this question was minimal, but in 2013 Yitang Zhang stunned the mathematical world by announcing that he could prove that infinitely many pairs of primes differ by at most 70 million. His paper has since been accepted for publication by the leading pure mathematics journal *Annals of Mathematics*. This may sound feeble compared to the twin prime conjecture, but it was the first time anyone had shown that infinitely many primes differ by some fixed amount. If 70 million could somehow be reduced to 2, this would resolve the twin primes conjecture.

Today's mathematicians increasingly use the Internet to join forces on problems, and Terence Tao orchestrated a collaborative effort to reduce the figure of 70 million to something much smaller. He did this within the framework of the Polymath project, a system set up to facilitate this kind of work. As mathematicians gained a better understanding of Zhang's methods, the number tumbled. James Maynard reduced the figure of 70 million to 600 (indeed to 12 if another conjecture called the Elliott–Halberstam Conjecture is assumed, see page 278). By the end of 2013 new ideas by Maynard had reduced it to 270.

Not yet 2, but a lot closer than 70 million.

The Odd Goldbach Conjecture

The second prime mystery to be resolved (probably!) goes back to 1742, when the German amateur mathematician Christian Goldbach wrote a letter to Leonhard Euler containing several observations about primes. One was: "Every integer greater than 2 can be written as the sum of three primes." Euler recalled a previous conversation in which Goldbach had made a related conjecture: "Every even integer is the sum of two primes."

With the convention then prevailing, that 1 is prime, this

Goldbach's letter to Euler, stating that if a number is a sum of two primes then it is a sum of any number of primes (up to the size of the number concerned). In the margin is the conjecture that every number greater than 2 is a sum of three primes. Goldbach defined 1 to be prime, which is not the modern convention.

statement implies the first one, because any number can be written as either $n+1$ or $n+2$ where n is even. If n is the sum of two primes, the original number is the sum of three primes. Euler said "I regard [the second statement] as a completely certain theorem, although I cannot prove it." That pretty much sums up its status today.

However, we no longer deem 1 to be prime, as discussed on page 90. So nowadays we split Goldbach's conjectures into two different ones:

The *even Goldbach conjecture* states:

> Every even integer greater than 2 is the sum of two primes.

The *odd Goldbach conjecture* is:

> Every odd integer greater than 5 is the sum of three primes.

The even conjecture implies the odd one, but not conversely.

Over the years, various mathematicians made progress on these questions. Perhaps the strongest result on the even Goldbach conjecture is that of Chen Jing-Run, who proved in 1973 that every sufficiently large even integer is the sum of a prime and a semiprime (either a prime or a product of two primes).

In 1995 the French mathematician Olivier Ramaré proved that every even number is a sum of at most six primes, and every odd number is a sum of at most seven primes. There was a growing belief among the experts that the odd Goldbach conjecture was within reach, and they were right: in 2013 Harald Helfgott claimed a proof using related methods. His result is still being checked by experts, but seems to be holding up well under scrutiny. It implies that every even number is the sum of at most four primes (if n is even then $n-3$ is odd, hence a sum of three primes $q+r+s$, so $n=3+q+r+s$ is a sum of four primes). This is close to the even Goldbach conjecture, but it seems unlikely that this can be proved in full using current methods. So there's still some way to go.

● ●

Prime Number Mysteries

Mathematics has mysteries of its own, and mathematicians who try to resolve them are like detectives. They seek out clues, make logical deductions, and look for proof that they are correct. As in Soames's cases, the most important step is knowing how to get started—what line of thinking might lead to progress. In many cases, *we still don't know*. This may sound like an admission of ignorance, and indeed it is. But it is also a statement that new mathematics still remains to be found, so the subject has not run dry. The primes are a rich source of plausible things that we don't actually know are true. Here are a few of them. In all cases p_n denotes the nth prime.

Agoh–Giuga Conjecture

A number p is prime if and only if the numerator of $pB_{p-1}+1$ is divisible by p, where B_k is the kth Bernoulli number [Takashi Agoh 1990]. Look those up on the Internet if you really want to know: the first few are $B_0 = 1$, $B_1 = \frac{1}{2}$, $B_2 = \frac{1}{6}$, $B_3 = 0$, $B_4 = -\frac{1}{30}$, $B_5 = 0$, $B_6 = \frac{1}{42}$, $B_7 = 0$, $B_8 = -\frac{1}{30}$.

Equivalently: A number p is prime if and only if

$$[1^{p-1} + 2^{p-1} + 3^{p-1} + \cdots + (p-1)^{p-1}] + 1$$

is divisible by p [Giuseppe Giuca 1950].

A counterexample, if it exists, must have at least 13,800 digits [David Borwein, Jonathan Borwein, Peter Borwein, and Roland Girgensohn 1996].

Andrica's Conjecture

If p_n is the nth prime, then

$$\sqrt{p_{n+1}} - \sqrt{p_n} < 1$$

[Dorin Andrica 1986].

Imran Ghory has used data on the largest prime gaps to confirm the conjecture for n up to 1.3002×10^{16}. The figure plots $\sqrt{p_{n+1}} - \sqrt{p_n}$ against n for the first 200 primes. The number 1 is at the top of the vertical axis, and all of the spikes shown are lower than that. They seem to shrink as the n gets bigger, but for all we know, there might be a huge spike poking out above 1 for some very large n. In order for the conjecture to be false, there would have to be an extremely large gap between two very large consecutive primes. This seems highly unlikely, but can't yet be ruled out.

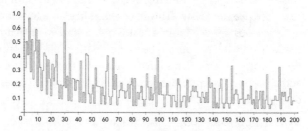

Plot of $\sqrt{p_{n+1}} - \sqrt{p_n}$ against n for the first 200 primes

Artin's Conjecture on Primitive Roots

Any integer a, other than -1 or a perfect square, is a primitive root modulo infinitely many primes. That is, every number between 1 and $p-1$ is a power of a minus a multiple of p. There are specific formulas for the proportion of such primes as their size becomes large [Emil Artin 1927].

Brocard's Conjecture

When $n > 1$ there are at least four primes between p_n^2 and p_{n+1}^2 [Henri Brocard 1904]. This is expected to be true; indeed much stronger statements ought to be true.

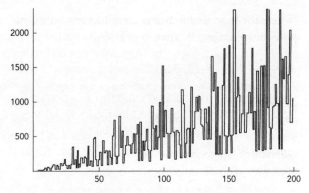

Number of primes between p_n^2 and p_{n+1}^2 plotted against n [Eric W. Weisstein, 'Brocard's Conjecture', from *MathWorld*—A Wolfram Web Resource: http://mathworld.wolfram.com/BrocardsConjecture.html]

Cramér's Conjecture

The gap $p_{n+1} - p_n$ between consecutive primes is no greater than a constant multiple of $(\log p_n)^2$ when n becomes large [Harald Cramér 1936].

Cramér proved a similar statement replacing $(\log p_n)^2$ by $\sqrt{p_n}\log p_n$, assuming the Riemann Hypothesis—perhaps the most important unsolved problem in mathematics, see *Cabinet* page 215.

Firoozbakht's Conjecture

The value of $p_n^{1/n}$ is strictly decreasing [Farideh Firoozbakht 1982]. That is, $p_n^{1/n} > p_{n+1}^{1/(n+1)}$ for all n. It is true for all primes up to 4×10^{18}.

First Hardy–Littlewood Conjecture

Let $\pi_2(x)$ denote the number of primes $p \leq x$ such that $p+2$ is also prime. Define the twin prime constant

$$C_2 = \prod_{p \geqslant 3} \frac{p(p-2)}{(p-1)^2} \approx 0.66016$$

(where the Π symbol indicates a product extending over all prime numbers $p \geq 3$). Then the conjecture is that

$$\pi_2(n) \sim 2C_2 \frac{n}{(\log n)^2}$$

where $\sim$ means that the ratio tends to 1 as n becomes arbitrarily large [Godfrey Harold Hardy and John Edensor Littlewood 1923].

There is a second Hardy–Littlewood conjecture (below).

Gilbreath's Conjecture

Start with the primes

2, 3, 5, 7, 11, 13, 17, 19, 23, 29, 31, ...

Compute the differences between consecutive terms:

1, 2, 2, 4, 2, 4, 2, 4, 6, 2, ...

Repeat the same calculation for the new sequence, ignoring signs, and continue. The first five sequences are

1, 0, 2, 2, 2, 2, 2, 2, 4, ...
1, 2, 0, 0, 0, 0, 0, 2, ...
1, 2, 0, 0, 0, 0, 2, ...
1, 2, 0, 0, 0, 2, ...
1, 2, 0, 0, 2, ...

Gilbreath and Proth conjectured that the first term in each sequence is always 1, no matter how many times the process is continued [Norman Gilbreath 1958, François Proth 1878].

Andrew Odlyzko verified the conjecture for the first 3.4×10^{11} sequences in 1993.

Goldbach's Conjecture for Even Numbers

Every even integer greater than 2 can be expressed as the sum of two primes [Christian Goldbach 1742].

T. Oliveira e Silva has verified the conjecture by computer for $n \leq 1.609 \times 10^{18}$.

Grimm's Conjecture

To each element of a set of consecutive composite numbers one can assign a distinct prime that divides it [C.A. Grimm 1969].

For example, if the composite numbers are 32, 33, 34, 35, 36, then one such assignment is 2, 11, 17, 5, 3.

Landau's Fourth Problem

In 1912 Edmund Landau listed four basic problems about primes, now known as Landau's problems. The first three are Goldbach's conjecture (above), the twin prime conjecture (below), and Legendre's conjecture (below). The fourth is: are there infinitely many primes p such that $p-1$ is a perfect square? That is, $p = x^2 + 1$ for integer x.

The first few such primes are 2, 5, 17, 37, 101, 197, 257, 401, 577, 677, 1297, 1601, 2917, 3137, 4357, 5477, 7057, 8101, 8837, 12101, 13457, 14401, and 15377. A larger example, by no means the largest, is

$$p = 1,524,157,875,323,883,675,049,535,156,256,668,194,$$
$$500,533,455,762,536,198,787,501,905,199,875,019,052,$$
$$101$$

$$x =$$
$$1,234,567,890,123456789,012,345,678,901,234,567,890$$

In 1997 John Friedlander and Henryk Iwaniec proved that infinitely many primes are of the form $x^2 + y^4$ for integer x, y. The first few are 2, 5, 17, 37, 41, 97, 101, 137, 181, 197, 241, 257, 277, 281, 337, 401, and 457. Iwaniec has proved that infinitely many numbers of the form $x^2 + 1$ have at most two prime factors. Close, but no banana.

Legendre's Conjecture

Adrien-Marie Legendre conjectured that there is a prime between n^2 and $(n+1)^2$ for every positive n. This would follow from Andrica's conjecture (above) and Oppermann's conjecture (below). Cramér's conjecture (above) implies that Legendre's conjecture is true for all sufficiently large numbers. It is known to be true up to 10^{18}.

Lemoine's Conjecture or Levy's Conjecture

All odd integers greater than 5 can be represented as the sum of an odd prime and twice a prime [Émile Lemoine 1894, Hyman Levy 1963].

The conjecture has been verified up to 10^9 by D. Corbitt.

Mersenne Conjectures

In 1644 Marin Mersenne stated that the numbers $2^n - 1$ are prime for $n = 2, 3, 5, 7, 13, 17, 19, 31, 67, 127$ and 257, and composite for all other positive integers $n < 257$. Eventually it was shown that Mersenne had made five errors: $n = 67$ and 257 give composite numbers and $n = 61, 89, 107$ give primes. Mersenne's conjecture led to the New Mersenne Conjecture and the Lenstra–Pomerance–Wagstaff Conjecture, which follow.

New Mersenne Conjecture or Bateman-Selfridge-Wagstaff Conjecture

For any odd p, if any two of the following conditions hold, then so does the third:

1 $p = 2^k \pm 1$ or $p = 4^k \pm 3$ for some natural number k.

2 $2^p - 1$ is prime (a Mersenne prime).

3 $(2^p + 1)/3$ is prime (a Wagstaff prime).

[Paul Bateman, John Selfridge, and Samuel Wagstaff Jr 1989]

Lenstra–Pomerance–Wagstaff Conjecture

There is an infinite number of Mersenne primes, and the number of Mersenne primes less than x is approximately $e^\gamma \log \log x / \log 2$ where γ is Euler's constant, roughly 0.577 [Hendrik Lenstra, Carl Pomerance, and Wagstaff, unpublished].

Oppermann's Conjecture

For any integer $n > 1$, there is at least one prime between $n(n - 1)$ and n^2, and at least another prime between n^2 and $n(n + 1)$ [Ludvig Henrik Ferdinand Oppermann 1882].

Polignac's Conjecture

For any positive even n, there are infinitely many cases of two consecutive primes with difference n [Alphonse de Polignac 1849].

For $n = 2$, this is the twin prime conjecture (see below). For $n = 4$, it says there are infinitely many *cousin primes* $(p, p + 4)$. For $n = 6$, it says there are infinitely many *sexy primes* $(p, p + 6)$ with no prime between p and $p + 6$.

Redmond–Sun Conjecture

Every interval $[x^m, y^n]$ (that is, the set of numbers running from x^m to y^n) contains at least one prime, except for $[2^3, 3^2]$, $[5^2, 3^3]$, $[2^5, 6^2]$, $[11^2, 5^3]$, $[3^7, 13^3]$, $[5^5, 56^2]$, $[181^2, 2^{15}]$, $[43^3, 282^2]$, $[46^3, 312^2]$, $[22434^2, 55^5]$ [Stephen Redmond and Zhi-Wei Sun 2006].

The conjecture has been verified for all intervals $[x^m, y^n]$ below 10^{12}.

Second Hardy–Littlewood Conjecture

If $\pi(x)$ is the number of primes up to and including x then

$$\pi(x + y) \leq \pi(x) + \pi(y)$$

for $x, y \geq 2$ [Godfrey Harold Hardy and John Littlewood 1923].

There are technical reasons for expecting this to be false, but the first violation is likely to occur for very large values of x, probably greater than 1.5×10^{174} but less than 2.2×10^{1198}.

Twin Prime Conjecture

There are infinitely many primes p such that $p + 2$ is also prime.

On 25 December 2011, PrimeGrid, a 'distributed computing project' that makes use of spare time on volunteers' computers, announced the largest pair of twin primes currently known:

$$3{,}756{,}801{,}695{,}685 \times 2^{666{,}669} \pm 1$$

These numbers have 200,700 digits.

There are 808,675,888,577,436 twin prime pairs below 10^{18}.

The Optimal Pyramid

Think of ancient Egypt and you think of pyramids. Especially the Great Pyramid of Khufu at Giza, the largest of them all, flanked by the slightly smaller pyramid of Khafre and the relative minnow of Menkaure. The remains of over 36 major Egyptian pyramids and hundreds of smaller ones are known; they range from huge almost complete ones to holes in the ground containing a few bits of stone from the burial chamber—or less.

Left: The Giza pyramids: *From back to front*: Great Pyramid of Khufu, Khafre, Menkaure, and three Queens' pyramids. Perspective makes those behind look smaller than they really are. *Right*: The bent pyramid.

An enormous amount has been written about the shapes, sizes, and orientations of pyramids. Most of it is speculative, using numerical relationships to construct ambitious chains of argument. The Great Pyramid is especially susceptible to this treatment, and it has been variously linked to the golden number, π, and even the speed of light. There are so many problems with this kind of reasoning that it's hard to take it seriously: the data are often inaccurate anyway, and with so many measurements to play with it's easy to come up with anything you want.

One of the best sources for data is Mark Lehner's *The Complete Pyramids*. Among other things, it lists the slopes of the faces: the angles between the planes formed by a triangular face and the square base. A few examples:

pyramid	angle
Khufu	51° 50′ 40″
Khafre	53° 10′
Menkaure	51° 20′ 25″
Bent	54° 27′ 44″ (lower), 43° 22′ (upper)
Red	43° 22′
Black	57° 15′ 50″

You can find more extensive data at

http://en.wikipedia.org/wiki/List_of_Egyptian_pyramids

Two observations spring to mind. The first is that stating some of these angles to the nearest second of arc (and others to the nearest minute) is unwise. The black pyramid of Amenemhat III at Dashur has a base of 105 metres and a height of 75metres. A difference of 1 second of arc in the slope corresponds to a change in height of 1 millimetre. Admittedly, there are traces of the edges of the base, and some fragments of casing stones may have survived, but given what remains of the pyramid, you'd be hard put to estimate the original slope to within 5° of the true figure.

What remains of the black pyramid of Amenemhat III

The other is that although the slopes vary somewhat—within a single monument in the case of the bent pyramid—they have a tendency to cluster around 54° or so. Why?

In 1979 R.H. Macmillan [Pyramids and pavements: some thoughts from Cairo, *Mathematical Gazette* 63 (December 1979)

251–255] started from the well-attested fact that the pyramid builders used expensive casing stones on the outside of their pyramids, such as white Tura limestone or granite. Inside they used cheaper materials: low-grade Mokattam limestone, mudbrick, and rubble. So it makes sense to reduce the amount of stone casing. What shape should a pyramid be if the pharaoh wants the biggest possible monument for a given cost of casing stones? That is: which angle of slope maximises the volume for a fixed area of the four triangular faces?

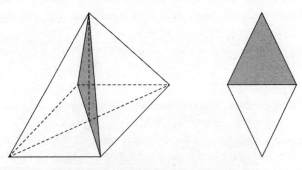

Left: Slicing a pyramid. *Right*: Maximising the area of an isosceles triangle, or equivalently a rhombus with given side.

This is a nice exercise in calculus, but it can also be solved geometrically using a clever trick. Slice the pyramid in half vertically through a diagonal of the base (shaded triangle). This yields an isosceles triangle. The volume of the half-pyramid is proportional to the area of this triangle, and the areas of the sloping faces of the half-pyramid are proportional to the lengths of the corresponding sides of the triangle. So the problem is equivalent to finding the isosceles triangle of maximal area when the lengths of the two equal sides are fixed.

Reflecting the triangle in its base, this is equivalent to finding the rhombus of maximal area with a given length of side. The answer is a square (oriented in a diamond position). So the angles of each triangular section of this kind are 90° at the top, and 45°

for the two base angles. It follows by basic trigonometry that the angle of slope of a face of the pyramid is

$$\arctan \sqrt{2} = 54° \ 44'$$

which is close to the average figure for actual pyramids.

Macmillan makes no claims about what this tells us about building pyramids; his main point is that it's a neat exercise in geometry. However, the Moscow mathematical papyrus contains a rule for finding the volume of a truncated pyramid (one with the top cut off), and a problem showing that the Egyptians understood similarity. It also explains how to find the height of a pyramid from its base and slope. Moreover, both this papyrus and the Rhind mathematical papyrus explain how to find the area of a triangle. So the Egyptian mathematicians could have solved Macmillan's problem.

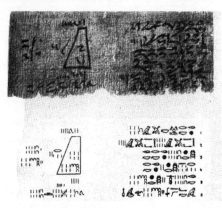

Moscow mathematical papyrus Problem 14: finding the volume of a truncated pyramid

In the absence of a papyrus containing that exact calculation, there's no convincing reason to suppose that they did. We don't have any evidence that they were interested in optimising the shape of their pyramids. Even if they were, they could have estimated the shape using clay models. Or made an educated guess. Or the shape could have evolved towards the least expensive one: builders and pharaohs are like that. Alternatively, the angle of slope might have been determined by engineering considerations: the bent pyramid is widely believed to be the shape it is because halfway up it started to collapse, so the slope was reduced. That said, this little piece of pyramid mathematics has more going for it than connections to the speed of light.

• •

The Sign of One: Part Two

From the Memoirs of Dr Watsup

Soames began striding up and down the room like a man possessed. Inwardly I gave a loud "huzzah!"—for I could see that he was hooked. Now I would reel him in out of the dark depression into which he had fallen, and rid myself of Bolivian funeral dirges to boot.

"We must be more systematic, Watsup!" he declared.

"In what way, Soames?"

"In a more systematic way, Watsup." The ensuing silence led him to be less obscure. "We must list small numbers that can be derived from just *two* 1's. By putting those together we can— well, you'll see in a minute, I'm sure."

So Soames wrote down:

$$0 = 1 - 1$$
$$1 = 1 \times 1$$
$$2 = 1 + 1$$
$$3 = \sqrt{1/.\dot{1}}$$
$$4 = \left\lceil \sqrt{1/.1} \right\rceil$$
$$5 = \left\lceil \sqrt{\left\lceil \sqrt{1/.1} \right\rceil !} \right\rceil$$
$$6 = (\sqrt{1/.\dot{1}})!$$

at which point Soames ground to a halt.

"I admit that 7 and 8 constitute temporary *lacunae*," he said. "Never mind, allow me to continue:

$$9 = 1/.\dot{1}$$
$$10 = 1/.1$$
$$11 = 11"$$

"I confess I do not yet—"

"Be assured, Watsup, you will. Suppose, for argument's sake, that we can express 7 and 8 using two 1's. Then we would have control of all numbers from 0 to 11. So, given any number *n* expressible using two 1's, we would be able to express everything between $n-11$ and $n+11$ using *four* 1's—merely by subtracting or adding the expressions on my systematic list."

"Ah, I see now," I said.

"You usually do, once I've told you," he replied acerbically.

"Then let me contribute a new thought to show that I have understood! Since we know how to express 24 using two 1's, for example as $\left\lceil \sqrt{1/.1} \right\rceil!$, we immediately express every number from $24-11$ to $24+11$ using four 1's. That is, the range from 13 to 35, inclusive."

"Exactly! I think we need not write those expressions down."

"No, Aha! We can get further! Look:

$$36 = \left(\left(\sqrt{1/.\dot{1}} \right)! \right) \times \left(\left(\sqrt{1/.\dot{1}} \right)! \right)"$$

"Yes," he replied. "But before your excitement carries you into realms untold, I will remind you that we do not yet have expressions for 7 and 8 using only two 1's."

I looked suitably crestfallen. But then a wild thought struck me. "Soames?" I asked tentatively.

"Yes?"

"Factorials make numbers bigger?"

He nodded, irritably.

"And square roots make them smaller?"

"Agreed. Get to the point, man!"

"And floors and ceilings round things off to whole numbers?"

I could see the realisation dawning on his face. "Stout fellow, Watsup! Yes, I see it now. We know, for example, how to express 24 using two 1's. Therefore we can also express 24! using two 1's, and that is"—his eyebrows narrowed—"620,448,401,733,239, 439,360,000. Whose square root is"—his face reddened as he carried out the mental arithmetic—"887,516.46, whose square root is 942.08, whose square root is 30.69."

"So we can express 30 and 31 using only two 1's," I said. "Namely:

$$30 = \left\lfloor \sqrt{}\sqrt{}\sqrt{}\left(\left(\left\lceil \sqrt{1/.1} \right\rceil !\right)!\right) \right\rfloor$$
$$31 = \left\lceil \sqrt{}\sqrt{}\sqrt{}\left(\left(\left\lceil \sqrt{1/.1} \right\rceil !\right)!\right) \right\rceil \text{''}$$

None of which helps us express 7 and 8 with two 1's, of course, but if we could do that, we could extend the range of numbers to $31 + 11$, which is 42. All of which argues, as you so cogently put it, Soames, that we should be systematic. I propose that we now investigate repeated square roots of factorials of numbers that we can express with two 1's."

"Agreed! And it is immediately obvious," said Soames, "that such an expression for 7 immediately yields one for 8."

"Uh—is it?"

"Naturally. Since 7! = 5040, whose square root is 70.99, whose square root is 8.42, we deduce that

$$8 = \left\lfloor \sqrt{}\sqrt{}(7!) \right\rfloor \text{''}$$

So, not for the first time in human history, the key to the mystery is the number 7!" By which, dear reader, he was placing an emphasis on the number 7, not referring to its factorial. Please pay attention, I have explained this before.

Soames's brow furrowed. "I can do it using a double factorial."

"You mean a factorial of a factorial?"

"No."

"A subfactorial? You've not yet explained—"

"No. The double factorial is a trifle obscure; it is

$$n!! = n \times (n-2) \times (n-4) \times \cdots \times 4 \times 2$$

when n is even, and

$$n!! = n \times (n-2) \times (n-4) \times \cdots \times 3 \times 1$$

when n is odd. So, for example,

$$6!! = 6 \times 4 \times 2 = 48$$

whose square root is 6.92, whose ceiling is 7."

I meekly wrote down

$$7 = \left\lceil \sqrt{\left(\left(\sqrt{1/.\dot{1}!}\right)!!\right)} \right\rceil$$

But Soames remained dissatisfied.

"The problem, Watsup, is that by introducing ever more obscure arithmetical functions, we could express any number whatsoever with ease. We might use Peano, for instance."

I objected vociferously. "Soames, you know that our landlady complains incessantly about your clarinet. She would never permit a piano!"

"Giuseppe Peano was an Italian logician, Watsup."

"To be honest, that might not make much difference. I'm not sure that Mrs Soapsuds would—"

"Quiet! In Peano's axiomatisation of arithmetic, the *successor* of any integer n is

$$s(n) = n + 1$$

So Peano could write

$$1 = 1$$
$$2 = s(1)$$
$$3 = s(s(1))$$
$$4 = s(s(s(1)))$$
$$5 = s(s(s(s(1))))$$

and the pattern continues indefinitely. Every integer would be expressible using just *one* 1. Or, for that matter, just one 0, since 1 = $s(0)$. It is too trivial, Watsup."

Can you find a way to write 7 using only two 1's and not using anything more esoteric than the functions Soames and Watsup employed before they started arguing about double factorials and successors? See page 279 for the answer.

Soames and Watsup have not finished yet. The Sign of One continues on page 115.

Initial Confusion

R.H. Bing

R.H. Bing was an American mathematician, born in Texas, who specialised in what became known as Texan topology. What does the R.H. stand for? Well, his father was Rupert Henry, but his mother felt that this sounded too British for Texas, so when he was christened she cut it down to the bare initials. So R.H. stands for R.H., nothing more. This caused a certain amount of puzzlement, but nothing too serious, until Bing applied for a visa to visit somewhere or other. When asked his name, anticipating the usual reaction, he said it was "R-only H-only Bing".

He received a visa made out to Ronly Honly Bing.

Euclid's Doodle

This is a mathematical mystery that was solved over two thousand years ago, and used to be taught in schools, but not any more—for sensible reasons. However, it's worth knowing, because it's far more efficient than the method that's usually taught instead. And it links up to all sorts of important bits of mathematics at higher levels.

People like to doodle. You see them on the phone, engaged in conversation, idly filling in all the o's on a page of the newspaper with a ballpoint pen. Or drawing wiggly lines that go round and round like irregular spirals. This sense of the word 'doodle', which originally meant a fool, seems to have been introduced by the screenwriter Robert Riskin in the 1936 comedy movie *Mr Deeds Goes to Town*; Mr Deeds refers to 'doodle' as scribbles that can help people think.

If a mathematician doodled—and most do—they might well get round to drawing a rectangle. What can you do to a rectangle? You can fill it in, you can draw spiral-like curves round the edge...or you can cut off a square from one end to make a smaller rectangle. Then it's only natural, and typical of the doodling mentality, to do the same again.

What happens? You might care to try a few rectangles before reading on.

OK, here we go. I started with a long, thin rectangle, and here's what happened.

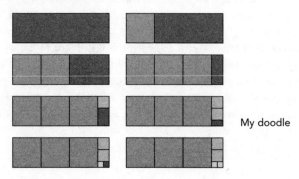

My doodle

Eventually I got to a small square, and ran out of rectangle.

Does this always happen? Does every rectangle eventually get gobbled up? Now *that's* a good question for a mathematician to think about.

What size was my rectangle? Well, the final picture shows that:

- The combined sides of two little squares equal the side of a medium square.
- The combined sides of two medium squares and one little one make the side of a big one, and also give one side of the rectangle.
- The combined sides of three big squares and a medium one make the other side of the rectangle.

If the small square has side 1 unit, then the medium one has side 2 and the large one has side $2 \times 2 + 1 = 5$. The short side of the rectangle is 5, and the long side is $3 \times 5 + 2 = 17$. So I started with a 17×5 rectangle.

That's interesting: by looking at the way the squares fit together, I can work out the shape of my rectangle. A more subtle implication is: if the process stops, the sides of the original rectangle are both *integer* multiples of the same thing: the side of the last square removed. In other words, the ratio of the two sides is of the form p/q for integers p and q. Which makes it a rational number.

This idea is completely general: if the doodle stops, the ratio of the sides of the rectangle is a rational number. In fact, the converse is also true: if the ratio of the sides of the rectangle is a rational number, the doodle stops. So doodles that stop correspond precisely to 'rational rectangles'.

To see why, let's look more closely at the numbers. The pictures in effect tell us this:

$$17 - 5 = 12$$
$$12 - 5 = 7$$
$$7 - 5 = 2$$

Now we have a 5×2 rectangle left and we have to move to the medium square

$$5 - 2 = 3$$
$$3 - 2 = 1$$

Now we have a 2×1 rectangle left and we have to move to a small square

$$2 - 1 = 1$$
$$1 - 1 = 0$$

Stop! And it *must* stop, because the integers involved are positive, and they are getting smaller at each stage. They must do, we're subtracting from them or leaving them alone. Now, a sequence of positive integers can't decrease forever. For example, if you start with a million and then decrease, you have to stop after at most a million subtractions.

More compactly, the doodle tells us that

17 divided by 5 gives 3 with remainder 2

5 divided by 2 gives 2 with remainder 1

2 divided by 1 goes exactly with zero remainder

and the process stops once the remainder is zero.

Euclid used his doodle to solve a problem in arithmetic: given two integers, calculate their highest common factor. This is the largest integer that divides both exactly; it is usually abbreviated to hcf. Another phrase is greatest common divisor, gcd. For instance, if the numbers are 4,500 and 840, the hcf is 120.

The way I was taught to do this at school is to factorise the two numbers into primes, and see what factors they have in common. For instance, suppose we want the hcf of 68 and 20. Factorise into primes:

$$68 = 2^2 \times 17 \qquad 20 = 2^2 \times 5$$

The hcf is $2^2 = 4$.

This method is limited to numbers that are small enough to be factorised quickly into primes. It's hopelessly inefficient for larger numbers. The ancient Greeks knew a more efficient method, a procedure that they gave the fancy name *anthyphairesis*. In this case it goes like this:

> 68 divided by 20 gives 3 with remainder 8
> 20 divided by 8 gives 2 with remainder 4
> 8 divided by 4 goes exactly with zero remainder
> Stop!

This is the same as the previous calculation with 17 and 5, but now all the numbers are four times the size (except for how many times each divides into the other, which stays the same). If you do the doodle with a 68×20 rectangle you get the same pictures as before, but the final small square is 4×4, not 1×1.

The technical name is Euclid's algorithm. An algorithm is a recipe for a calculation. Euclid put this one in his *Elements*, and he used it as the basis of his theory of prime numbers. In symbols, the doodle goes like this. Take two positive integers $m \leq n$. Start with the pair (m, n) and replace it by $(m, n-m)$ in numerical order, smallest first: that is, transform

$$(m, n) \rightarrow (\min(m, n - m), \max(m, n - m))$$

where min and max are the minimum and maximum respectively. Repeat. At each stage the largest number in the pair gets smaller, so eventually the process stops with a pair $(0, h)$, say. Then h is the required hcf. The proof is easy: any factor of both m and n is also a factor of both m and $n-m$, and conversely. So at each step the hcf remains the same.

This method is genuinely efficient: you can use it by hand for really big numbers. To prove it, here's a question for you. Find the hcf of 44,758,272,401 and 13,164,197,765.

See page 279 for the answer.

Euclidean Efficiency

How efficient is Euclid's algorithm?

Chopping off one square at a time is simpler for theoretical purposes, but the more compact form, in terms of division with remainder, is the best one to use in practice. This telescopes all cuts using a fixed size of square into a single operation.

Most of the computational effort occurs in the division step, so we can estimate the efficiency of the algorithm by counting how many times this step gets used. The first person to investigate this question was A.A.-L. Reynaud, and in 1811 he proved that the number of division steps is at most m, the smaller of the two numbers. This is a very poor estimate, and he later got it down to $m/2 + 2$, not much better. In 1841 P.-J.-E. Finck reduced the estimate to $2 \log_2 m + 1$, which is proportional to the number of decimal digits in m. In 1844 Gabriel Lamé proved that the number of division steps is at most five times the number of decimal digits of m. So even for two numbers with 100 digits, the algorithm gets the answer in no more than 500 steps. In general, you can't do it that quickly using prime factors.

What's the worst-case scenario? Lamé proved that the algorithm runs most slowly when m and n are consecutive members of the Fibonacci sequence

$$1 \quad 1 \quad 2 \quad 3 \quad 5 \quad 8 \quad 13 \quad 21 \quad 34 \quad 55 \quad 89 \quad \ldots$$

in which each number is the sum of the previous two. For these numbers, exactly *one* square gets chopped off each time. For instance, with $m = 34$, $n = 55$, we get

       55 divided by 34 gives 1 with remainder 21
       34 divided by 21 gives 1 with remainder 13
       21 divided by 13 gives 1 with remainder 8
       13 divided by 8 gives 1 with remainder 5
       8 divided by 5 gives 1 with remainder 3
       5 divided by 3 gives 1 with remainder 2
       3 divided by 2 gives 1 with remainder 1
       2 divided by 1 goes exactly.

It's an unusually long calculation for such small numbers.

Mathematicians have also analysed the average number of division steps. With fixed n, the number of division steps averaged over all smaller m is approximately

$$\frac{12}{\pi^2} \log 2 \log n + C$$

where C, called *Porter's constant*, is

$$-\frac{1}{2} + \frac{6 \log 2}{\pi^2} (4\gamma - 24\pi^2 \zeta'(2) + 3 \log 2 - 2) = 1.467$$

Here $\zeta'(2)$ is the derivative of Riemann's zeta function evaluated at 2, and γ is Euler's constant, about 0.577. It would be hard to find a sensible problem leading to a more comprehensive selection of mathematical constants in one formula. The ratio of this formula to the exact answer tends to 1 as n gets larger.

123456789 Times X

Sometimes very simple ideas lead to mysterious results. Try multiplying 123456789 by 1, 2, 3, 4, 5, 6, 7, 8, and 9. What do you notice? When does it go wrong?

See page 279 for the answers. For an extension, see page 145.

The Sign of One: Part Three

From the Memoirs of Dr Watsup

Heaps of paper covered in arcane scribbles were sprouting like mushrooms from every flat surface in Soames's lodgings. This, you understand, was not unusual; Mrs Soapsuds often berated him about his deep litter filing system, to no avail. But on this occasion the scribbles were sums.

"I can obtain 8 using two 1's, without involving a hypothetical expression for 7," I announced. "In fact,

$$8 = \lfloor \sqrt{\sqrt{\sqrt{(11!)}}} \rfloor$$

But for the life of me I cannot derive 7."

"That one does seem tricky," Soames agreed. "But your result leads to progress in other ways:

$$14 = \lfloor \sqrt{\sqrt{(8!)}} \rfloor$$
$$15 = \lceil \sqrt{\sqrt{(8!)}} \rceil$$

where of course we substitute your expression for 8 where necessary. I could write it out in full—"

"No, no, Soames, I am already convinced!"

"But now we have two more *lacunae* at 12 and 13. However, Watsup, I suspect these problems are related. Let me see ... Well,

$$32 = \lfloor \sqrt{\sqrt{\sqrt{(15!)}}} \rfloor$$

and we already have 15 using only two 1's. Then

$$12 = \lfloor \sqrt{\sqrt{\sqrt{\sqrt{\sqrt{(32!)}}}}} \rfloor$$
$$13 = \lceil \sqrt{\sqrt{\sqrt{\sqrt{\sqrt{(32!)}}}}} \rceil$$

and further

$$16 = \lfloor \sqrt{\sqrt{\sqrt{(13!)}}} \rfloor$$
$$17 = \lceil \sqrt{\sqrt{\sqrt{(13!)}}} \rceil$$

and finally

$$7 = \lceil \sqrt{\sqrt{\sqrt{\sqrt{(16!)}}}} \rceil$$

which resolves the issue entirely satisfactorily. So substituting in turn for the various numbers, we find that

$$7 =$$
$$\lceil \sqrt{\sqrt{\sqrt{((\lfloor \sqrt{\sqrt{\sqrt{\sqrt{((\lceil \sqrt{\sqrt{\sqrt{\sqrt{\sqrt{((\lfloor \sqrt{\sqrt{\sqrt{(((\lceil \sqrt{\sqrt{((\lfloor \sqrt{\sqrt{(11!)}} \rfloor)!)}} \rceil)!)}}} \rfloor)!)}}}}} \rceil)!)}}}} \rfloor)!)}}} \rceil$$

I am mortified not to have seen that immediately."

"Is that the *simplest* solution, Soames?" said I, gulping. "I hope not!"

"I have no idea. Perhaps some ingenious person could do better. It is hard to be sure with these matters. I am sure they will inform us by telegram if they can better our own feeble efforts."

"Anway," I said, "if we can express any integer n with two 1's, we can now express the entire range from $n-17$ to $n+17$."

"Exactly, Watsup. Our quest becomes simpler by the moment. All we need is a series of numbers, each exceeding the previous one by no more than 35, so that their ranges abut or overlap. That will enable us to reach the largest such number plus 17."

"Which means—" I began—

"That we should be *systematic*!"

"Quite."

"We had already reached . . . remind me, Watsup. Consult those extensive notes of yours."

I delved into several piles of documents, eventually finding my notebook beneath a stuffed skunk. "We had reached 32, Soames, if we include your incidental remark earlier when seeking 7."

"And, of course,

$$33 = \lceil \sqrt{\sqrt{\sqrt{(15!)}}} \rceil$$"

said he. "Very good. So ideally we need to express 68, 103, 138, and so on, with two 1's. But we can use smaller numbers if those are more convenient. Provided that the increase is at most 35 from one to the next."

Several hours of intense computation, and more heaps of paper, led to a short but vital list:

$$71 = \lceil \sqrt{(7!)} \rceil \quad 79 = \lfloor \sqrt{\sqrt{(8!)}} \rfloor \quad 80 = \lceil \sqrt{\sqrt{(8!)}} \rceil \quad 120 = 5!$$

but little more.

"Perhaps I was too hasty in dismissing double factorials, Watsup."

"Very likely, Soames."

Soames nodded and wrote

$$105 = 7!!$$

Then, in a sudden fit of inspiration, he added

$$19 = \lfloor \sqrt{(8!!)} \rfloor$$
$$20 = \lceil \sqrt{(8!!)} \rceil$$

exclaiming "If we can find a way to write 18 with two 1's, then we extend the range surrounding any integer n expressible using two 1's: we can then deal with $n-20$ to $n+20$." He paused for breath, and added "Failing that, the only gaps would be $n-18$ and $n+18$, which perhaps we could derive in other ways."

"I think it is time we took stock," I said. I perused our scribbled notes. "It seems to me that we had expressed all numbers from 1 to 33 using four 1's. Then

$$43 = \lfloor \sqrt{\sqrt{(10!)}} \rfloor$$
$$44 = \lceil \sqrt{\sqrt{(10!)}} \rceil$$

need only two 1's, so we immediately fill in everything between 26 and 61. There is a gap at 62 (because that is $44 + 18$ and we are stuck on 18 if only two 1's are allowed) but we can do 63 and 64. Now, based on 80, we can continue to 97. Then we are stuck on 98, but 99 and 100 can be achieved."

"More easily, in point of fact," said Soames:

$$99 = 11/.\dot{1} \times 1$$
$$100 = 1/(.1 \times .1)$$
$$101 = 1/(.1 \times .1) + 1$$

"So we are complete up to 100," I said, "with the exceptions of 62 and 98."

"But 98 is taken care of by 105, along with all numbers up to 122," said Soames.

"Oh, I had forgotten we could do 105 with two 1's."

"And since $120 = 5!$, also expressible using two 1's, we can reach 137. Indeed, we also get 139 and 140."

"So the only gaps up to 140 are 62 and 138," I said.

"So it seems," said Soames. "I wonder if those gaps can be filled by some other method?"

Can you find a way to write 62 and 138 with four 1's, not using anything more esoteric than the functions Soames and Watsup have used so far? See page 280 for the answer.

Soames and Watsup still haven't finished. But the end is nigh: The Sign of One concludes on page 126.

See page 280 for the answer. The Sign of One concludes on page 126.

. . . .

Taxicab Numbers

Srinivasa Ramanujan

Srinivasa Ramanujan was a self-taught Indian mathematician with an amazing talent for formulas—usually very strange ones, yet having their own kind of beauty. He was brought to England by Cambridge mathematicians Godfrey Harold Hardy and John Edensor Littlewood in 1914. By 1919 he was terminally ill with lung disease, and he died in India in 1920. Hardy wrote:

> I remember once going to see him when he was lying ill at Putney. I had ridden in taxi-cab No. 1729, and remarked that the number seemed to be rather a dull one, and that I hoped it was not an unfavourable omen. "No," he replied, "it is a very interesting number; it is the smallest number expressible as the sum of two [positive] cubes in two different ways."

The observation that

$$1729 = 1^3 + 12^3 = 9^3 + 10^3$$

was first published by Bernard Frénicle de Bessy in 1657. If negative cubes are permitted then the smallest such number is

$$91 = 6^3 + (-5)^3 = 4^3 + 3^3$$

Number theorists have generalised the concept. The nth taxicab number, $Ta(n)$, is the smallest number that can be expressed as a sum of two positive cubes in n or more distinct ways.

In 1979 Hardy and E.M. Wright proved that some numbers can be expressed as the sum of an arbitrarily large number of positive cubes, so $Ta(n)$ exists for all n. However, to date only the first six are known:

$$Ta(1) = 2 = 1^3 + 1^3$$
$$Ta(2) = 1729 = 1^3 + 12^3 = 9^3 + 10^3$$
$$Ta(3) = 87539319 = 167^3 + 436^3 = 228^3 + 423^3$$
$$= 255^3 + 414^3$$
$$Ta(4) = 6963472309248$$
$$= 2421^3 + 19083^3 = 5436^3 + 18948^3$$
$$= 10200^3 + 18072^3 = 13322^3 + 166308^3$$
$$Ta(5) = 48988659276962496$$
$$= 38787^3 + 365757^3 = 107839^3 + 362753^3$$
$$= 205292^3 + 342952^3$$
$$= 221424^3 + 336588^3 = 231518^3 + 331954^3$$
$$Ta(6) = 24153319581254312065344$$
$$= 582162^3 + 28906206^3 = 3064173^3 + 28894803^3$$
$$= 8519281^3 + 28657487^3 = 16218068^3 + 27093208^3$$
$$= 17492496^3 + 26590452^3 = 18289922^3 + 26224366^3$$

$Ta(3)$ was discovered by John Leech in 1957. $Ta(4)$ was found by E. Rosenstiel, J.A. Dardis and C.R. Rosenstiel in 1991. $Ta(5)$ was found by J.A. Dardis in 1994 and confirmed by David Wilson

in 1999. In 2003 C.S. Calude, E. Calude, and M.J. Dinneen
established that the number stated above is probably Ta(6), and
in 2008 Uwe Hollerbach announced a proof.

The Wave of Translation

John Scott Russell

Mathematical research on horseback?

Why not? Inspiration strikes whenever it does. You don't get
to choose.

In 1834, John Scott Russell, a Scottish civil engineer and
naval architect, was riding his horse alongside a canal, when he
noticed something remarkable:

> I was observing the motion of a boat which was rapidly drawn
> along a narrow channel by a pair of horses, when the boat
> suddenly stopped—not so the mass of water in the channel
> which it had put in motion; it accumulated round the prow
> of the vessel in a state of violent agitation, then suddenly
> leaving it behind, rolled forward with great velocity,
> assuming the form of a large solitary elevation, a rounded,

smooth and well-defined heap of water, which continued its course along the channel apparently without change of form or diminution of speed. I followed it on horseback, and overtook it still rolling on at a rate of some eight or nine miles an hour, preserving its original figure some thirty feet long and a foot to a foot and a half in height. Its height gradually diminished, and after a chase of one or two miles I lost it in the windings of the channel. Such, in the month of August 1834, was my first chance interview with that singular and beautiful phenomenon which I have called the Wave of Translation.

He was intrigued by this discovery, because normally individual waves spread out as they travel, or break like surf on a beach. He constructed a wave tank at his home, and carried out a series of experiments. These revealed that this kind of wave is very stable, and it can travel a long way without changing shape. Waves of different sizes travel at different speeds. If one such wave catches up with another one, it emerges in front after a more complicated interaction. And a big wave in shallow water will divide into two: a medium one and a small one.

These discoveries puzzled the physicists of the time, because their current understanding of fluid flow could not explain it. In fact, George Airy, a distinguished astronomer, and George Stokes, the leading authority on fluid dynamics, had trouble believing it. We now know that Russell was right. In appropriate circumstances, nonlinear effects, beyond the scope of the mathematics of his era, counteract the tendency of a wave to spread out because the speed of the wave depends on its frequency. These effects were first understood around 1870 by Lord Rayleigh and Joseph Boussinesq.

In 1895 Diederik Korteweg and Gustav de Vries came up with the Korteweg–de Vries equation, which included such effects, and showed that it has solitary wave solutions. Similar results were derived for other equations of mathematical physics, and the phenomenon acquired a new name: solitons. A series of major discoveries led Peter Lax to formulate very general conditions for equations to have soliton solutions, and explained the 'tunnelling' effect. It is mathematically quite different from the way that shallow waves interact by superposing their shapes, like two sets of ripples crossing each other on a pond, which is a direct consequence of the mathematical form of the wave equation. Soliton-like behaviour occurs in many areas of science, from DNA to fibre optics. This has led to a wide range of new phenomena with names like breathers, kinks, and oscillons.

There's also a tantalising idea that has not yet been made to work. Fundamental particles in quantum mechanics somehow combine two apparently different characteristics. Like most things at a quantum level, they are waves, yet they hang together in a particle-like clump. Physicists have tried to find equations that respect the structure of quantum mechanics but allow solitons to exist. The closest they have come so far is an equation that produces an instanton: this can be interpreted as a particle that has a very short lifetime, winking into existence from nowhere and disappearing immediately.

Riddle of the Sands

Barchan dunes. *Left*: Paracas National Park, Peru. *Right*: Hellespontus region, from Mars Reconnaissance Orbiter.

Sand dunes form a variety of patterns: linear, transverse, parabolic, ... One of the most intriguing is the barchan, or crescent, dune. The name comes from Turkestan, and is said to have been introduced into geology in 1881 by the Russian naturalist Alexander von Middendorf. Barchans can be found in Egypt, Namibia, Peru,... and even on Mars. They are crescent-shaped, come in a range of sizes, and they *move*. They form swarms, interact with each other, break up and join together. In recent years mathematical modelling has provided a lot of insight into their shapes and behaviour, but many mysteries remain.

Dunes are formed by the interaction of wind and sand grains. The rounded end of a barchan faces into the prevailing wind, which pushes the sand up the front of the dune and round the sides, where it forms two trailing arms that give it its characteristic crescent shape. At the top of the dune, the sand tumbles over and is sucked down the 'slip face' between the arms. A large vortex of rotating air called a separation bubble scours out the space between the arms.

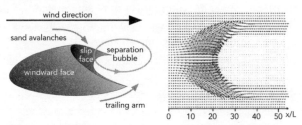

Left: Schematic of a barchan and the separation bubble. *Right*: Simulation by Barbara Horvat of the motion of sand grains, computed from a mathematical model.

Barchans behave like solitons (see previous item), although technically they differ in some respects. As the wind blows them along, the small dunes travel faster than the big ones. If a small dune catches up with a big one, it appears to be absorbed, but after a time the big barchan spits out a small one, almost as through the small one had tunnelled through the big one. The small one then heads off, faster than the lumbering beast behind it.

In their paper, Veit Schwämmle and Hans Herrmann comment on the similarities and differences between barchan

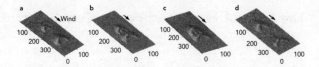

Simulated collision of a small barchan and a larger one by Veit Schwämmle and Hans Herrmann. (a) Time 0: small dune behind big one. (b) After 0.48 years: small dune catches big one and they collide. (c) After 0.63 years: dunes mixed together. (d) After 1.42 years: small dune in front of big one.

collisions and solitons. The figure shows what happens if the two dunes are of similar sizes. Initially (a) the smaller dune is behind the larger one but moving faster. It hits the back of the larger one (b) and climbs up its windward face, but gets stuck part way (c). Then the front separates to form a smaller dune (d).

For some combinations of heights, the emerging dune is bigger than the smaller one was originally, whereas for others it is smaller. This behaviour differs from that of solitons, where both waves end up the same size as they began. However, there is an intermediate range of height combinations for which the dunes retain their sizes and volumes *exactly*. In these cases they behave like solitons.

If the small dune is a lot smaller than the big one, it just gets gobbled up and a single, larger, barchan forms. If the difference in heights is moderate, the collision can result in 'breeding': two small barchans appear at the tips of the horns of the larger one, and head off in front of it. Real barchans do all these things. The dynamics of barchan dunes is richer than that of conventional solitons.

● ●

Eskimo π

Why is π only 3 in the Arctic?

Everything shrinks in the cold.

● ●

The Sign of One: Part Four—Concluded

From the Memoirs of Dr Watsup

"Well, this is a pretty pickle," I muttered.

"A gherkin, I believe," said Soames, plucking the vinegar-soaked vegetable from the jar and consuming it with relish.

I put the relish back in the pantry, along with the pickle jar.

"We do have the option," Soames remarked, "of multiplying numbers by 3, 9, or 10 using only one extra 1. We merely divide by $\sqrt{.\dot{1}}$, $.\dot{1}$, or $.$"

"Then I have it!" I cried:

$$62 = 63 - 1 = 7 \times 9 - 1 = 7/.\dot{1} - 1$$

recalling that we can express 7 using only two 1's—indeed, in at least two different ways."

"Leaving only 138 to perplex us."

"It is 3×46," I mused. "Can we achieve 46 using just three 1's? Then we could divide that by $\sqrt{.\dot{1}}$ as you suggested."

A systematic search of floors and ceilings of repeated square roots of factorials led us to an unexpected discovery: it is possible to achieve 46 with only *two* 1's. I present only the solution: the route to its discovery involved many dead ends and failures. Start with a representation of 7 using two 1's, for example,

$$7 = $$
$$\lceil \sqrt{\sqrt{\sqrt{((\lfloor\sqrt{\sqrt{\sqrt{((\lceil\sqrt{\sqrt{\sqrt{\sqrt{((\lfloor\sqrt{\sqrt{\sqrt{((\lceil\sqrt{\sqrt{((\lfloor\sqrt{\sqrt{(11!)}}\rfloor)!})}}\rceil)!})}}\rfloor)!})}}\rceil)!})}}\rfloor)!})}}\rceil$$

Then observe that

$$70 = \lfloor\sqrt{7!}\rfloor$$
$$37 = \lceil\sqrt{\sqrt{\sqrt{\sqrt{\sqrt{\sqrt{70!}}}}}}\rceil$$
$$23 = \lceil\sqrt{\sqrt{\sqrt{\sqrt{\sqrt{37!}}}}}\rceil$$
$$26 = \lceil\sqrt{\sqrt{\sqrt{\sqrt{23!}}}}\rceil$$
$$46 = \lfloor\sqrt{\sqrt{\sqrt{\sqrt{26!}}}}\rfloor$$
$$138 = 46/\sqrt{.\dot{1}}$$

Working backwards and substituting the formulas for the numbers, the end result expresses 138 with just three 1's.

"Shall I write it out explicitly, Soames?"

"Good heavens no! Anyone who wishes to see the full formula can do that for themselves."

Buoyed by this unanticipated success, I wanted to continue even further with our list. But Soames merely shrugged. "Perhaps the problem merits further computation. Perhaps not."

A thought struck me. "Might we prove that every number can be obtained with four 1's—perhaps even fewer—by iterating floors and ceilings of repeated square roots of factorials?"

"It is a plausible conjecture, Watsup, but to be frank I see no route to a proof, and the strain of so much mental arithmetic is beginning to tell."

He was sinking back into depression. Desperately I suggested "You could try logarithms, Soames."

"I thought of those right at the start, Watsup. You may be surprised to hear that using nothing more than logarithms the exponential function, and the ceiling function, *any* positive integer can be expressed using only *one* 1."

"No, no, I meant using logarithms as a computational aid, not in the formula—" But Soames ignored my protestations.

"Recall that the exponential function is

$$\exp(x) = e^x \text{ where } e = 2.71828\ldots$$

and its inverse function is the natural logarithm

$$\log(x) = \text{whichever } y \text{ satisfies } \exp(y) = x$$

Is that not so, Watsup?" I affirmed that to the best of my knowledge it was.

"Then we merely observe that

$$n + 1 = \lceil \log(\lceil \exp(n) \rceil) \rceil$$

whose proof is straightforward."

I gawped at him, but managed to get out a strangled "Of course, Soames."

"Then we iterate:

$$1 = 1$$
$$2 = \lceil \log(\lceil \exp(1) \rceil) \rceil$$
$$3 = \lceil \log(\lceil \exp(\lceil \log(\lceil \exp(1) \rceil) \rceil) \rceil) \rceil$$
$$4 = \lceil \log(\lceil \exp(\lceil \log(\lceil \exp(\lceil \log(\lceil \exp(1) \rceil) \rceil) \rceil) \rceil) \rceil) \rceil$$

and—"

I hastily grasped his writing hand. "Yes, Soames, I understand. It is a thinly disguised variation on the Peano method, which we rejected earlier because of its triviality."

"So the game is no longer afoot, Watsup, if exponentials and logarithms are permitted."

I concurred, with some sadness, for immediately he took up his clarinet and commenced to play a rhythmless atonal composition by some obscure Eastern European composer. It sounded like a cat caught in a mangle. A tone-deaf cat. With a sore throat.

His black mood would now be unshakeable.

Here Ends The Sign of One.

Except that I still haven't told you what a subfactorial is. That comes next.

• •

Seriously Deranged

Time to explain subfactorials.

Suppose that *n* people each own a hat. They all pick up one hat and put it on. In how many different ways can this be done so that no one gets their own hat? Such an assignment is called a *derangement*.

For example, if there are three people—Alexandra, Bethany, and Charlotte, say—then their hats can be assigned in six ways:

ABC ACB BAC BCA CAB CBA

For ABC and ACB, Alexandra gets her own hat, so these are not derangements. For BAC, Charlotte gets her own hat. For CBA,

Bethany gets her own hat. That leaves two derangements: BCA and CAB.

With four people—suppose Deirdre joins the group—there are 24 arrangements:

~~ABCD~~	~~ABDC~~	~~ACBD~~	~~ACDB~~	~~ADBC~~	~~ADCB~~
~~BACD~~	BADC	~~BCAD~~	BCDA	BDAC	~~BDCA~~
~~CABD~~	CADB	~~CBAD~~	~~CBDA~~	CDAB	CDBA
DABC	~~DACB~~	~~DBAC~~	~~DBCA~~	DCAB	DCBA

but 15 of them (crossed out) assign someone their own hat. (Remove anything with A in the first position, B in the second, C in the third, or D in the fourth.) So there are 9 derangements.

The number of derangements of n objects is the subfactorial (denoted by $!n$ or $n_{\text{¡}}$). This has a number of definitions. The simplest is probably

$$!n = \left\lfloor \frac{n!}{e} + \frac{1}{2} \right\rfloor$$

Its values begin

$!1 = 0$	$!2 = 1$	$!3 = 2$	$!4 = 9$
$!5 = 44$	$!6 = 265$	$!7 = 1{,}854$	$!8 = 14{,}833$
$!9 = 133{,}496$	$!10 = 1{,}334{,}961$		

* *

Tossing a Fair Coin Isn't Fair

Tossing a coin

The fair coin is a staple of probability theory, equally likely to land heads or tails. It is generally considered to be the epitome of randomness. On the other hand, a coin can be modelled as a simple mechanical system, and as such its motion is completely determined by the initial conditions when it is tossed—mainly the vertical velocity, the initial rate at which it spins, and the axis of spin. This makes the motion non-random. So where does the randomness in a coin toss come from? I'll come back to that after describing a related discovery.

Persi Diaconis, Susan Holmes, and Richard Montgomery have shown that tossing a 'fair' coin isn't actually fair. There is a small but definite bias: when a coin is tossed, it is slightly more likely to land the same way up as its initial orientation on your thumb. In fact, the chance of it doing that is approximately 51%. Their analysis assumes that the coin doesn't bounce when it hits the ground, which is reasonable for grass, or when it is caught in the hand, but not if it lands on a wooden tabletop.

The 51% bias becomes statistically significant only after about 250,000 flips. It arises because the axis about which the coin is spinning might not be horizontal. As an extreme case, suppose that the axis is at right angles to the coin, so the coin always remains horizontal as it spins, like a potter's wheel. In this case, it will always land the same way up as it was to begin with, a 100% chance of not flipping over. The other extreme case is when the spin axis is horizontal, and the coin flips end over end. Although in principle the final state is then determined by the upward velocity and rate of spin when the coin leaves the hand, even small errors in specifying these numbers imply that the coin lands the same way up as it started only 50% of the time. With this type of toss, the predetermined state of a mechanical coin is randomised by small errors.

Usually, the spin axis is in neither of these extreme locations, but somewhere in between, and close to the horizontal. So there is a slight bias in favour of landing the same way up. Detailed calculations lead to the 51% figure. Experiments with a coin-tossing machine confirm this figure reasonably well.

In practice a real coin toss is random, with a 50% chance of heads or tails, for none of these reasons. It is random because the

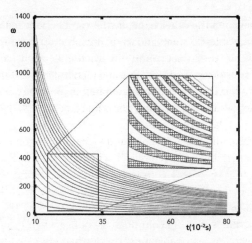

How heads (white) and tails (shaded) vary with the initial rate of rotation (vertical axis) and the time spent in the air (horizontal axis) when the spin axis is horizontal. The head/tail stripes become very closely spaced when the spin rate is high.

initial orientation of the coin, when it's sitting on the thumb, is random. In the long run, the coin starts heads up half the time, and tails up half the time. This removes the 51% bias because the initial state is not known when the coin is tossed.

See page 280 for further information.

● ●

Playing Poker by Post

Suppose that Alice and Bob—the traditional participants in any cryptographic exchange—want to play poker, specifically, five-card stud. But Alice is in Alice Springs, Australia, while Bob is in Bobbington, a town in Staffordshire, England. Can they perhaps mail each other the cards? The main problem is *dealing* the cards, a 'hand' of five to each player. How can both players be sure that each has a hand from the same pack, without the other knowing their hand?

If Bob just mails Alice five cards, she can't be sure that he hasn't seen her cards; moreover, when Bob plays cards from what allegedly is his hand, she can't be sure whether he really has only five cards to work with, or whether the remainder of the pack is available to him, and he is only pretending to use a fixed hand of five cards, dealt before the game started.

If this hand arrived in the post, you could probably be sure the dealer wasn't cheating. But for most hands, how can you tell?

Surprisingly, it *is* possible to play a card game like poker by mail, or over the phone, or over the Internet, without any danger that either player is cheating. Alice and Bob can use number theory to create codes, and resort to a complicated series of exchanges. Their method is known as a zero knowledge protocol, a way to convince someone that you possess a specific item of knowledge *without telling them what it is*. For example, you could convince an online banking system that you know the security code on the back of your credit card, without conveying any useful information about the code itself.

Hotels often lock guests' valuables in a safety deposit box in the reception area. To ensure security, each box has two keys: one kept by the manager and the other by the guest. *Both* keys are needed to open the box. Alice and Bob can use a similar idea:

1 Alice locks a card in each of 52 boxes, using padlocks whose codes only she knows. She mails the lot to Bob.

2 Bob (who cannot unlock the boxes to see what cards are inside) picks five boxes and mails them back to Alice. She unlocks them and receives her five cards.

3　Bob chooses another five boxes, and puts an extra padlock on each. He knows the codes for unlocking these, but Alice does not. He mails the boxes to Alice.

4　Alice removes *her* padlocks from these boxes, and mails them back to Bob. He can now open them to receive his five cards.

After these preliminaries, the game can start. Cards are revealed by mailing them to the other player. To prove no one cheated, they can unlock all the boxes after the end of the game.

Alice and Bob convert this idea to mathematics by extracting the essential features. They represent the cards by an agreed set of 52 numbers. Alice's padlocks correspond to a code A, known only to Alice. This is a function, a mathematical rule, that changes a card number c into another number Ac. (I'm taking liberties with the notation by not writing $A(c)$, in order to avoid talking about 'composing' functions.) Alice also knows the inverse code A^{-1}, which decodes Ac back into c. That is,

$$A^{-1}Ac = c$$

Bob doesn't know A or A^{-1}.

Similarly, Bob's padlocks correspond to codes B and B^{-1}, known only to Bob, such that

$$B^{-1}Bc = c$$

With these preliminaries, the method corresponds to the padlock procedure like this:

1　Alice sends all 52 numbers $Ac_1, \ldots, Ac_{52}$ to Bob. He has no idea which cards these correspond to; in effect, Alice has shuffled the pack.

2　Bob 'deals' five cards to Alice and five to himself. He sends Alice her cards. To simplify the notation, consider just one of these, and call it Ac. Alice can find c by applying A^{-1}, so she knows what cards are in her hand.

3　Bob needs to find out what his five cards are, but only Alice knows how to work that out. But he can't send his cards to Alice because then she will know what they are. So for each card Ad in his hand, he applies his own code B to get BAd, and sends *that* to Alice.

4   Alice can again apply A^{-1} to 'remove her padlock', but this time there's a snag: the result is

$$A^{-1}BAd$$

In ordinary algebra we could swap A^{-1} and B round to get

$$BA^{-1}Ad$$

which equals

$$Bd$$

Then Alice could send that back to Bob, who would then apply B^{-1} to find d.

However, functions can't be swapped like this. For example, if $Ac = c+1$ (so $A^{-1}c = c-1$) and $Bc = c^2$, then

$$A^{-1}Bc = Bc-1 = c^2-1$$

whereas

$$BA^{-1}c = (A^{-1}c)^2 = (c-1)^2 = c^2-2c+1$$

which is *different*.

The way round this obstacle is to avoid that sort of function, and set up the codes so that $A^{-1}B = BA^{-1}$. In this case, A and B are said to *commute*, because a little algebra turns this into the equivalent condition $AB = BA$. Notice that in the physical method, Alice's and Bob's padlocks do indeed commute. They can be applied in either order, and the result is the same: a box with two padlocks.

Alice and Bob can therefore play poker by post if they can set up two *commuting* codes A and B, so that the decoding algorithm A^{-1} is known only to Alice, and B^{-1} is known only to Bob.

Bob and Alice agree on a large prime number p, which can be public knowledge. They agree 52 numbers $c_1, \ldots, c_{52} \pmod{p}$ to represent the cards.

Alice picks some number a between 1 and $p-2$, and defines her code function A by

$$Ac = c^a \pmod{p}$$

Using basic number theory, the inverse (decoding) function is of the form

$$A^{-1}c = c^{a'} \pmod{p}$$

for a number a' that she can compute. Alice keeps both a and a' secret.

Bob similarly chooses a number b and defines his code function B by

$$Bc = c^b \pmod{p}$$

with inverse

$$B^{-1}c = c^{b'} \pmod{p}$$

for a number b' that he can compute. He keeps both b and b' secret.

The code functions A and B commute, because

$$ABc = A(c^b) = (c^b)^a = c^{ba} = c^{ab} = (c^a)^b = B(c^a) = BAc$$

where all equations hold $\pmod{p}$. So Alice and Bob can use A and B as described.

• •

Eliminating the Impossible

From the Memoirs of Dr Watsup

"Watsup!"

"Uh—I'm not sure, Soames. What *is* up?"

"I'm calling your name, man, not asking a question! How many times have I told you not to bring copies of *The Strand* magazine into this house?"

"But—how—"

"You know my methods. You were tapping your fingers impatiently, as you do when waiting for me to go out. And your eyes kept flicking towards the newspaper rolled up in your coat pocket. Which is too thick for the *Daily Reporter*, despite what it says on the front page, so it must contain a magazine. Since you habitually conceal only one such from me, its identity was never in doubt."

"I'm sorry, Soames. I was just hoping to gain some comparative insights into investigative methods from the writings of the companion of the, er, charlatan living across the street from us."

"Pah! The man is a fraud! A mountebank who calls himself a detective!"

There are times when Soames can be overbearing. Indeed, few when he is not, now that I come to think of it. "I have occasionally sifted some useful hints from my exploited counterpart's outpourings of vapidity, Soames," I objected.

"Such as?" he asked, in an aggressive tone.

"I am impressed by his argument: 'When you have eliminated the impossible, then whatever remains, how ever improbable, must be—' "

"—wrong," Soames brusquely completed my sentence. "If what remains is truly improbable, then you have almost surely made some unstated assumption when declaring other explanations to be impossible."

Consistency is not one of Soames's virtues. "Well, perhaps, but—"

"No buts, Watsup!"

"But on other occasions you have agreed—"

"Pah! Reality is not improbable, Watsup. It may look that way, but its probability is 100%, *because it has happened*."

"Yes, technically, but—"

"A case in point. This morning, Watsup, while you were out buying that scurrilous rag, I received an unexpected visitor. The Duke of Bumbleforth."

"The toast of London," I said. "A noble man of unimpeachable probity, a role model for us all."

"Indeed. Yet he informed me . . . Well, there had, he said, been a dinner party at Bumbleforth Hall, at which the Earl of Maundering attempted to entertain the guests by arranging ten wine glasses in a row and filling the first five of them—like this."

He demonstrated with our own glasses, filling them with a rather acid Madeira that we had decided to throw out. "He then challenged the guests to rearrange the glasses so that they alternated full and empty."

"But that is easy," I began.

"If you move four glasses, yes. Interchange the second with the seventh and the fourth with the ninth. Like so." (See the figure below.) "However, the Earl's challenge was to obtain the same result by moving only *two* of the glasses."

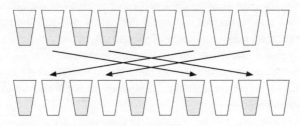

How to solve the puzzle in four moves

I pressed my fingers together in an attitude of deep thought, and after a moment I drew a rough sketch of the original and final arrangements. "But Soames: the four glasses you have named must all end up in different locations! So all four must move!"

He nodded. "So, Watsup, you have now eliminated the impossible."

"By Jove, yes I have, Soames! Incontrovertibly."

He began stuffing tobacco into his pipe. "So what do you conclude if I tell you that according to the Duke of Bumbleforth, after all of the guests had expressed similar opinions, the Earl of Maundering then demonstrated a solution."

"I—uh—"

"You are forced to conclude that the honourable Duke, a scion of the British Empire and a man of high nobility . . . is actually a base liar. For no solution exists, as you have proved."

My face fell. "It does seem—no, wait, perhaps *you* are not telling me the—"

"My dear doctor, I freely confess that I dissemble from time to time, always with your own best interests at heart, but not on this occasion. You have my word."

"Then...I am shocked at the Duke's behaviour."

"Come now, Watsup, Have faith in British character."

"The Earl cheated?"

"No, no, no. Nothing of the kind. You can do better than this. There may be another perfectly prosaic explanation that you have overlooked. In fact, I predict that you will shortly be telling me how childishly simple the answer is."

Soames then told me what Maundering had done.

"Why, how childishly sim—" I began. I stopped abruptly, and am forced on grounds of candidness to admit that I flushed a deep crimson.

What was Maundering's solution? See page 281 for the answer.

Mussel Power

It's an idyllic seaside scene: a quiet bay with waves breaking over the rocks, which are festooned with clumps of shellfish and seaweed. But those sedate static mussel beds are actually a hive of activity. To see it, you just have to speed up the flow of time. In time-lapse photography, the mussels are constantly on the move. They tether themselves to the rocks using special threads, secreted by their foot. By detaching some threads and adding others in new locations, the mussels can control their positions on the rocks. On the one hand (foot?), they like to stay close to other mussels because that way they are not so likely to be ripped off the rock by the waves. On the other, they get more food if there aren't other mussels nearby to compete with them. Faced with this dilemma, mussels do what any sensible organism would: they compromise. They arrange themselves so that they

have a lot of near neighbours but few distant ones. That is, they
congregate in patches. You can see the patches with the naked
eye, but not how they form.

Clump of blue mussels

In 2011 Monique de Jager and co-workers applied the
mathematics of random walks to deduce how the mussels'
clumping strategy might have evolved. A random walk is often
compared to a drunkard moving along a path: sometimes
forward, sometimes back, with no clear pattern. Going up a
dimension, a random walk in the plane is a series of steps, whose
lengths and directions are chosen randomly. Different rules for
the choices—different probability distributions for the lengths
and directions—lead to random walks with different properties.
In Brownian motion, the length is distributed in a bell curve,
close to one specific average step size. In a Lévy walk, the
probability of making a step is proportional to some fixed power
of its size, so many short steps are occasionally interrupted by a
much longer one.

Statistical analysis of observed step sizes show very clearly
that a Lévy walk fits what mussels on intertidal mudflats actually
do, whereas Brownian motion doesn't. This agrees with
ecological models, which demonstrate mathematically that a
Lévy walk disperses the mussels faster, opens up more new sites,

and avoids competition with other species of shellfish. This in turn suggests why this particular strategy may have evolved. Natural selection provides a feedback loop between movement strategies and the genetic instructions that cause them to be used. Individual mussels are more likely to survive if they employ strategies that increase their chances of obtaining food, and decrease their chances of being swept away by a wave.

De Jager's team used field observations of what mussels do and simulations of mathematical models of the evolutionary process. The simulations showed that Lévy walks are likely to evolve as a result of this population-level feedback, and the conditions for that strategy to be evolutionarily stable—that is, not susceptible to invasion by a mutant with a different strategy—predict that the power-law exponent should be 2. The field data indicate a value of 2.06.

The novel feature of mussel beds, in this context, is that the effectiveness of an individual's movement strategy depends on what all the other mussels are doing. Each mussel's strategy is determined by its own genetics, but the survival value of that strategy depends on the collective behaviour of the entire local population. So here we see how the environment—in the form of the other mussels—interacts with individual genetic 'choices' to produce pattern formation on the population level.

See page 281 for further information.

Proof That the World is Round

Most of us are aware that our planet is round—not an exact sphere, though: a bit flattened at the poles. It has enough bumps to turn it into a potato if you exaggerate the discrepancy from a spheroid by a factor of about 10,000. A few—very few—hardy souls persist in the belief that the world is flat, even though the ancient Greeks, 2,500 years ago, amassed evidence of its rotundity that convinced even medieval clerics, and more evidence has been piling up ever since. Belief in a flat Earth almost died out, but it was revived around 1883 with the founding of the Zetetic Society. This became the Flat Earth Society in 1956, and it is still active today. You can find it on the Internet and follow it on Facebook and Twitter.

There's an easy and virtually foolproof way to check for yourself that our planet can't be flat if the usual geometry of Euclid applies. It requires Internet access or a tolerant travel agent, but no other special apparatus, and it's not looking up the shape on Wikipedia. The method doesn't of itself show that the Earth is round, but a systematic and careful extension would be able to do just that. I'll discuss potential ways to deny this evidence in a moment. I don't claim there's no way out—if you're a flat-Earther there's *always* a way out. But in this instance, the standard ploys are even less convincing than usual. In any case, the argument makes a refreshing change from the usual scientific evidence for a round world.

I'm not thinking of satellite photos of a round planet—those are, of course, fakes. We all know that NASA never went to the Moon, it was all done in Hollywood, which proves they're fakes, so there. Nothing that relies on scientific measurement, either: those scientist types are well-known hoaxers, they even pretend they believe in evolution and global warming, both of which are leftie plots to stop clean-living, righteous-thinking people making the obscene sums of money that are their God-given right.

No, what I have in mind is commercial evidence: airline flight times. You can look these up on the Internet: make sure you use actual flights that exist, not flight-time calculators that assume a round Earth.

For commercial reasons all large passenger jets fly at about the same speed. If they didn't, other companies would get all the business from the slow ones. They fly the shortest route, subject to local regulations, for similar reasons. So we can use travel times as reasonably accurate estimates of distances. (To reduce the effects of wind, take a suitable average of flight times in both directions—in practice the usual arithmetic mean is good enough, but see page 281.) The surveyor's technique of triangulation, which constructs a network of triangles, can then be used to map out the locations of the airports concerned. For the purpose of showing that a flat Earth doesn't work, we can assume it's flat and see what that implies. Surveyors usually work with one initial distance, the baseline, and calculate everything else from the angles of the triangle, but we have the luxury of using actual distances (in units of aircraft-hours).

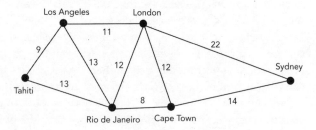

Flat-Earth airline map

The figure shows a triangulation based on six major airports. Give or take a little jiggling, this is the only planar arrangement that is a reasonable fit to the travel times. Start with London and add Cape Town distance 12 away. After that place Rio de Janeiro and Sydney. Their locations are unique, except that the entire map could be reflected left-right without changing any distances. That ambiguity doesn't matter, but you do have to confirm that Rio de Janeiro and Sydney are on opposite sides of the line from London to Cape Town. If they were on the same side, the flight time between them would be about 11 hours, but actually it's 18. You can add Los Angeles next, and finally locate Tahiti, again using extra timings to remove ambiguities.

Now we can use the hypothesis of a flat Earth to make a prediction. The distance from Tahiti to Sydney, measured from the map, is about 35 hours. (As it happens, the route via Rio and Cape Town is very close to a straight line, and the sum of the distances is 35.) So that's the *minimum* time it should take to travel by plane, not counting stops.

The actual figure is 8 hours. Even allowing for minor errors, the difference is way too large, and the hypothesis of a flat planet must be rejected. With a network with many more airports, and more precise figures, you could map out the basic shape of much of the planet very accurately—still in units of aircraft-hours. To set the scale you'd need to know how fast the planes fly, or make at least one distance measurement some other way.

United Nations logo: azimuthal equiangular projection from round Earth to flat disc

Now, every well-informed flat-Earther is aware of forms of words and non-standard physics that can 'explain' these results. Perhaps some kind of distortion field alters the geometry, so that the literal image of a plane with its usual measure of distance is wrong. This really does work: an azimuthal equiangular projection of the Earth from the north pole does just that, and you can transfer everything from a round Earth to a flat one, laws of physics included, using projection onto a flat disc. Provided you miss out the region around the south pole. The United Nations logo does just that, and has been used by the Flat Earth Society to 'prove' that its views are correct. However, this kind of change is trivial and meaningless. It's a logically equivalent model to a round Earth with conventional geometry. Mathematically it is just a less than candid way to admit 'it's not flat', within the orthodox meaning of that phrase. So altered metrics and similar excuses really don't hack it.

Effects of wind? Maybe there's a really high wind blowing from Tahiti to Sydney? It would have to blow at 750 mph, but worse than that: the straight line path between Tahiti and Sydney is very close to the routes Tahiti-Rio-Cape Town-Sydney that we've already taken into account. If you could get from Tahiti to Sydney really quickly by exploiting the wind, at least one of those routes is taking way too long.

The next line of defence would be the standard denialist ploy: it's all a huge conspiracy. Yes, but by whom? The times listed on Internet booking sites can't be far wrong, because millions of people travel by air every day and most of them would notice if the scheduled times were frequently wildly wrong. But the airlines might all be conspiring to fly more slowly than necessary on some routes, so that most of my map should shrink, making it possible to get from Tahiti to Sydney in a mere 14 hours. You have to divide by 4 or more to make this possible, so a conventional passenger jet could actually get from London to Sydney in 5 hours if the airline wasn't deliberately dawdling to convince us the planet is round.

Unlike allegations of scientists conspiring, which only make sense to people who don't know any scientists,* this one has a flaw that is pretty much fatal. It requires most airlines to lose huge amounts of money every day in wasted fuel, and to refrain from wiping out the competition by flying routes in less than half the time they currently take. A conspiracy to make the Earth appear round, using the airline-schedule metric, would require hundreds of private-sector companies to voluntarily throw away vast sums of *money*. Are you mad?

You can, of course, always fall back on that old standby: when all else fails, ignore the evidence.

. . . .

---

* I'm not referring to the proposition that scientists are mostly honest. I'm referring to the delight they all take in proving each other wrong, which among other things is how they get promoted. Massive conspiracies would still make no sense even if all scientists were crooks.

123456789 Times X Continued

There's no need to stop at 9 (see page 115). Try multiplying 123456789 by 10, 11, 12, and so on. What do you notice now?

See page 282 for the answer.

. . . .

The Price of Fame

Władysław Orlicz

Władysław Roman Orlicz, a Polish topologist, introduced what are now known as Orlicz spaces. These are highly technical concepts in functional analysis. One day his fame proved counterproductive. Like most of his compatriots, he lived in a very small apartment, and one day he applied to the city officials asking for a bigger one. The reply was: "We agree that your apartment is very small, but we must deny your claim since you have your own spaces."

. . . .

The Riddle of the Golden Rhombus

From the Memoirs of Dr Watsup

The spectacular success of our joint endeavours encouraged me once more to take up medical practice, and I arranged for a small surgery to be constructed in my house. But I took care to allow sufficient flexibility for occasions when Soames might require my services, with or without adequate notice. So when the telegram arrived, I handed my patient over to my locum Dr Jekyll, and summoned a cab to take me to 222B Baker Street.

When I arrived at Soames's lodgings I found him surrounded by pieces of paper. In his hands was a pair of scissors.

"A pretty puzzle," he remarked. "Merely a rectangular strip of paper, tied into a simple overhand knot. It is hard to imagine that a man's fate may depend upon it."

Knotted strip of paper

"Great Heavens, Soames! How could that be?"

"A nasty case of extortion, Watsup. The evidence hinges on the shape that is formed when the knot is pulled as tightly as possible, and flattened out. I suspect that it will turn out to be the symbol of a secret society, and if I can so prove, my case will be complete." He held the knot before my eyes. "So, Watsup, what shape will we see, eh?"

I quickly sketched an overhand knot in my notebook.

Overhand knot tied in a closed loop of string

"It is well known that when an overhand knot is tied in a closed loop, it has threefold symmetry," said I, feeling unusually astute. "I would therefore expect either a triangle or a hexagon to form."

"Let us try the experiment, then," said Soames. "And then we shall tackle the more difficult task of proving that our eyes do not deceive us."

What shape is the flattened knot? Try it. See page 282 for the answer and a proof.

A Powerful Arithmetic Sequence

An arithmetic sequence (a sequence of numbers with constant differences) is *powerful* if the second term is a square, the third term a cube, and so on. That is, the kth term is a kth power. (This imposes no condition on the first term since all numbers are first powers of themselves.) For example 5, 16, 27 has length 3 and common difference 11, and

$$5 = 5^1 \qquad 16 = 4^2 \qquad 27 = 3^3$$

A trivial way to get a powerful sequence of length n is to repeat n times the number $2^{n!}$. This is a first power, a square, a cube, and so on, up to an nth power. The common difference is zero.

In 2000 John Robertson proved that excluding sequences like this, where the same number repeats—common difference zero—the longest possible powerful sequence has five terms (length 5). See John P. Robertson, The maximum length of a powerful arithmetic progression, *American Mathematical Monthly* 107 (2000) 951. To obtain this sequence, start with the numbers 1, 9, 17, 25, 33, which form an arithmetic sequence with common difference 8, and multiply each of them by $3^{24}5^{30}11^{24}17^{20}$. The resulting numbers also form an arithmetic sequence, with common difference 8 times this number. They are:

(1)   10529630094750052867957659797284314695762718513 64140020404487941414117813110351565

(2)   94766670852750475811618938175558832261864466622 772601836403914727270603179931640625

(3)   179003711610750898755280216553833349827966214573 190380346876295004040002822875976565

(4)   263240752368751321698941494932107867394067962844 1035005101121985353529453277587890625

(5)   347477793126751744642602773310382384960169710950 166206733481020666658878326416015625

The common difference is

   842370407580004229436612783782745175661017481091 3120163235903531312942504882125000

If the five terms are a_1, a_2, a_3, a_4, a_5, then

  a_1 is the first power of itself (obviously)
  $a_2 = 307841957589849138828884412917083740234375^2$ is a
    square
  $a_3 = 5635779747116948576103515625^3$ is a cube
  $a_4 = 716288998461106640625^4$ is a fourth power
  $a_5 = 51072299355515625^5$ is a fifth power

Wow!

  (It's easier to check that the terms are the stated powers if you work with the prime factors.)

Why Do Guinness Bubbles Go Downwards?

Anyone who drinks dark stout, such as Guinness, will have seen something that appears to fly in the face of conventional physics. The bubbles in the beer move downwards. At least, they seem to. But bubbles are lighter than the surrounding fluid, so they experience a buoyancy force that pushes them *up*.

It's a genuine mystery, or at least it was until 2012 when a team of mathematicians solved it. Appropriately, they were Irish (or based in Ireland): William Lee, Eugene Benilov, and Cathal Cummins of the University of Limerick.

The same effect occurs in other liquids, but it's easier to see in stout because light-coloured bubbles show up more clearly against a dark beer. It is enhanced because stout bubbles contain nitrogen as well as the carbon dioxide that is found in all beers, and nitrogen bubbles are smaller and last longer.

Part of the answer is easy: we're only seeing the bubbles that are near the glass. Those out in the middle are hidden from view by the dark beer. So maybe some bubbles are going up, but others down. What that fails to explain is why *any* bubbles go down. They shouldn't.

Until a few years ago we didn't even know whether the whole thing was an optical illusion. One alternative explanation is that the effect is caused by density waves—regions where bubbles bunch up. The bubbles go upwards but the density waves go the other way. This kind of behaviour is common in waves. For example, the water in ocean waves does not travel along with the wave; it mostly goes round and round in roughly the same place. What moves is the location of the highest parts of the water. Admittedly, waves breaking on a beach do go up the beach; however, some of that is the effect of shallow water, and the water runs back down to the sea. If the water were travelling with the waves, it would have to pile up ever higher on the beach, which doesn't make much sense. Although the water doesn't go backwards to any significant extent, this familiar example shows the difference between where the water goes and where the waves go. Now do it with bubbles.

It's a fairly plausible theory, but in 2004 a group of Scottish scientists led by Andrew Alexander, working with colleagues in California, produced video footage proving that the bubbles really do move downwards. The team released its results on St Patrick's Day. They used a high-speed camera to slow down the movement and track individual bubbles. They found that the bubbles touching the walls of the glass tended to stick, so they couldn't move upwards. However, bubbles near the middle were free to rise, which caused the beer to flow up in the middle and down at the sides, dragging the bubbles with it.

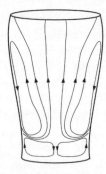

Flow of Guinness in a glass: down at the edges

The Irish team refined this explanation, showing that it's not caused by bubbles sticking to the walls. What makes it happen is the shape of the glass. Stout is usually drunk from a glass with curved sides, which is wider at the top than at the bottom. Using fluid dynamics calculations and experiment, the team found that when the bubbles near the wall rise, they go straight up, as you'd expect. But the wall slopes away from the vertical, so in effect the bubbles move away from the wall. The beer near the wall is therefore denser than that in the middle, so it tends to slide down the side of the glass, dragging the nearby fluid with it. So the beer circulates: upwards in the middle, downwards near the sides.

The bubbles are *always* going upwards relative to the beer, but at the edges the beer goes down faster than the bubbles go up, so the bubbles go down too. We see the bubbles, but we can't easily see the motion of the beer.

See page 286 for further information.

Random Harmonic Series

The infinite series

$$1 + \frac{1}{2} + \frac{1}{3} + \frac{1}{4} + \cdots + \frac{1}{n} + \cdots$$

is called the *harmonic series* by mathematicians. The name is loosely related to music, where the overtones of a vibrating string have wavelengths 1/2, 1/3, 1/4, and so on, of the string's fundamental wavelength. However, the series has no musical significance. It is known to be divergent, meaning that the sum up to n terms becomes as large as we wish if n is large enough. It diverges very slowly, but it does diverge. In fact, the first 2^n terms add up to more than $1 + n/2$. On the other hand, if we change the sign of every other term, we get the alternating harmonic series

$$1 - \frac{1}{2} + \frac{1}{3} - \frac{1}{4} + \cdots + (-1)^{n+1}\frac{1}{n} + \cdots$$

which converges. Its sum is log 2, which is about 0.693.

Byron Schmuland wondered what happens if successive signs are chosen at random, by tossing a fair coin and assigning a plus sign to 'heads' and a minus sign to 'tails'. He proved that, with probability one, this series converges (the harmonic series would correspond to tossing HHHHHH...forever, which has probability zero). However, the sum depends on the sequence of tosses.

The question now arises: what is the probability of getting a particular sum? The sum can be any real number, positive or negative, so the probability of getting any specific number is zero (this is generally the case for 'continuous random variables'). The way to deal with this is to introduce a probability distribution (or density) function. This determines the probability of getting a sum in any given *range* of values, say between two numbers *a* and *b*. This probability is the *area* under the distribution function between $x = a$ and $x = b$.

For harmonic series modified using random coin tosses, the probability distribution looks like the figure below. It's a bit like the familiar bell curve, or normal distribution, but the top looks flat. It has left–right symmetry, corresponding to swapping 'heads' and 'tails' on the symmetric coin.

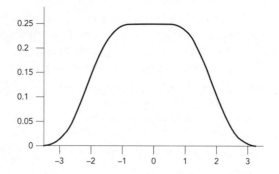

Probability distribution for random harmonic series

This problem is an object lesson in 'experimental mathematics', in which computer calculations are employed to suggest interesting conjectures. It looks as though the central peak is at height 0.25, that is, 1/4. It also looks as though the values of the function at −2 and +2 are 0.125, that is, 1/8. In 1995 Kent Morrison conjectured that both statements are true, but in 1998 he changed his mind, having investigated the conjectures in more detail. To ten decimal places, the density has value 0.2499150393 at $x = 0$, slightly less than 1/4. However, to ten decimal places the value at $x = 2$ is 0.1250000000, which still looks like 1/8. To 45 decimal places, however, the value turns out to be:

$$0.1249999999999999999999999999999999999999764$$

which differs from 1/8 by less than 10^{-42}.

Schmuland's paper [Random harmonic series, *American Mathematical Monthly* 110 (2003) 407–416] explains why this probability is so close to, but not exactly, 1/8. So here a very plausible conjecture from experimental evidence turns out to be *false*. This is why mathematicians insist on proofs, just as Hemlock Soames insists on evidence.

The Dogs That Fight in the Park

From the Memoirs of Dr Watsup

Taking my usual morning constitutional in Equilateral Park, just off Marylebone Road near the Dog and Triangle public house, I observed a curious incident, and on arriving to 222B Baker Street I could not restrain myself from sharing it with my colleague.

"Soames, I have just witnessed a curious—"

"Incident. You saw three dogs in the park," said he, not batting an eyelid.

"But how—of course! There is mud on my trousers, and the spatter patterns indicate—"

Soames chuckled. "No, Watsup, my deduction has another basis. It tells me not only that you saw three dogs in the park, but that they were fighting."

"So they were! But that was not the curious incident. It would have been curious had the dogs *not* fought."

"True. I must remember that remark, Watsup. Most apposite."

"What was curious was what preceded the fight. The dogs appeared simultaneously at the three corners of the park—"

"Which is an equilateral triangle whose sides are all 60 yards," Soames interjected.

"Uh, yes. At the moment they appeared, each dog faced the

next in a clockwise direction, and immediately began to run towards it."

"Each at the same speed of 4 yards per second."

"I bow to your judgement. As a result, all three dogs followed curved paths, and collided simultaneously at the centre of the park. In a flash, they were fighting, and I had to separate them."

"Whence the tears in your coat and trousers, and the teeth marks on your leg, which I see were inflicted by one red setter, one retriever, and one cross between a bulldog and an Irish wolfhound. With a lame front left leg."

"Ah."

"And wearing a red leather collar. With a bell on it. Which has rusted and no longer rings. Were you observant enough to notice how long it took the dogs to collide?"

"I neglected to look at my pocket-watch, Soames."

"Oh, come now, Watsup! You look, but you do not *see*. However, in this case the time can be deduced from the facts already established."

Assume the dogs are points. See page 286 for the answer.

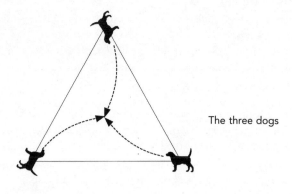

The three dogs

How Tall is That Tree?

There's an old forester's trick (the trick is old, not the forester) for estimating the height of a tree without climbing it or using surveying equipment. It can be a great icebreaker at garden parties if there's a suitable tree in the vicinity. I learned it from Toby Buckland, Digging deeper, *Amateur Gardening* (20 October 2012) page 59. Trousers are the recommended attire.

Stand a reasonable distance from the tree, with your back towards it. Bend over and look back at it through your legs. If you can't see the top, move away until you can. If you can see it easily, move closer until it's just visible. At that point, your distance from the base of the tree will be roughly equal to its height.

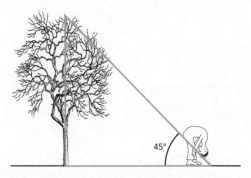

Estimating the height of a tree

The technique, if I may call it that, is a simple application of Euclidean geometry. It works because the angle at which most of us can look upwards through our legs is roughly 45°. So the line of sight to the top of the tree is the hypotenuse of an isosceles right-angled triangle, and the other two sides are equal.

Obviously the accuracy of the method depends on how flexible your body is, but it's not too far wrong for many of us. Buckland remarks: "Have a go, it's cheaper than yoga and offers a view on the world that most haven't enjoyed since childhood!"

Why Do My Friends Have More Friends Than I Do?

OMG! Everyone seems to have more friends than I do!

It happens on Facebook, it happens on Twitter. It happens on any social media website, and it happens in real life. It happens if you count business partners or sexual partners. It's a humbling experience when you start checking your friends to see how many friends *they* have. Not only do most of them have more than you do: on average, they *all* have more than you do.

Why are *you* so unpopular compared to everyone else? It's very worrying. But there's no need to be upset. Most people's friends have more friends than they do.

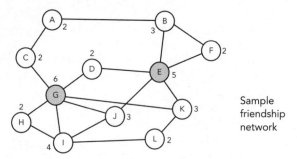

Sample friendship network

That probably sounds weird. Everyone in a given social network has *on average* the same number of friends; namely, the average – there is only one. Some have more, some have fewer, but the average is...the average. It then seems intuitively plausible that their friends will, on average, *also* have that very same number of friends. But is it true?

Let's try an example. It's not concocted to be unusual; it's the first one I drew. Most networks will behave similarly. The network (above) shows 12 people, with lines connecting the friends. (We assume all friendships work both ways. That's not always the case for social networks, but the effect still occurs even so.) Tabulate a few key figures:

person	number of friends	*their* nos. of friends	average of previous column
Alice	2	3, 2	**2.5**
Bob	3	2, 5, 2	3
Cleo	2	2, 6	**4**
Dion	2	5, 6	**5.5**
Ethel	5	3, 2, 2, 3, 3	2.6
Fred	2	3, 5	**4**
Gwen	6	2, 2, 2, 4, 3, 3	2.67
Hemlock	2	6, 4	**5**
Ivy	4	6, 2, 3, 2	3.25
John	3	5, 6, 4	**5**
Kate	3	5, 6, 2	**4.33**
Luke	2	4, 3	**3.5**

I've used boldface to mark the numbers in the final column that are bigger than the numbers in the second. These are the cases where person X's friends have, on average, more friends than person X has. *Eight out of twelve* numbers are boldface, and there's another where the two figures are equal.

The average of the numbers in column 2 is 3. That is, the average number of friends, across the entire social network, is 3. But the majority of entries in column 4 are bigger than that. So what's wrong with intuition here?

The answer is people like Ethel and Gwen who have an unusually large numbers of friends, here 5 and 6 respectively. Because of this, they get counted a lot more often when we're looking at how many friends people's *friends* have. And they then contribute more to the total in column 3, hence to the average. On the other hand, people with few friends show up much less often, and contribute less.

Your friends are not a typical sample. People with a lot of friends will be over-represented among them, because there's a greater chance that *you* are one of their friends. People with few friends will be under-represented. This effect skews the average towards a higher value.

You can see it happening in column 3 of the table. The

number 5 occurs five times in column 3, one for each of Ethel's friends; similarly 6 occurs six times in column 3, one for each of Gwen's friends. On the other hand, Alice's contribution to column 3 (not in her row, but when she occurs as a friend in other rows) is just two 2's: one from Bob and one from Cleo. So Ethel contributes 25, and Gwen a massive 36, whereas poor old Alice contributes just 4.

To them that hath, shall be given.

This does *not* happen for column 2: everyone contributes their fair share to the average, which is 3.

In fact the average of all the numbers in column 4 is 3.78, a lot bigger than 3. I probably ought to use a weighted average: add all the numbers in column 3 and divide by how many of them there are. This is 3.55, again bigger than 3.

I hope you feel happier now.

See page 287 for a proof.

See page 287 for a proof.

. . . .

Isn't Statistics Wonderful?

Statistically, 42 million alligator eggs are laid every year. Of those, only half hatch. Of those that hatch, three quarters are eaten by predators in the first month. Of the rest, only 5% are alive after a year, for one reason or another.

If it wasn't for statistics, we'd all be eaten by alligators!

. . . .

The Adventure of the Six Guests

From the Memoirs of Dr Watsup

It has long distressed me that Soames has a hearty dislike of dinner parties. He despises small talk and becomes uneasy in the company of women, especially attractive women like my friend Beatrix. But from time to time he is obliged to bite the bullet, grasp the nettle, plait the platitude, and attend social events that include the fairer sex. At these, he can be from one moment to the next taciturn, obnoxious, charming, garrulous, or some combination.

This particular occasion was a modest *tête-à-tête* with Aubrey and Beatrix Sheepshear (brother and sister) and Crispin and Dorinda Lambshank (husband and wife). I knew all four, of course; Beatrix is a delightful lady, unmarried, and without a current suitor, I firmly believe. Soames knew only me, which I feared might cause the worse side of his character to dominate, but I was hoping to widen his social circle. The Sheepshears and Lambshanks had not met, except for the men, who belonged to the same club.

All this became clear to Soames when the guests arrived, and soon we were sitting together. Soames's presence made the conversation spasmodic, so I ventured to pour some glasses of a modest but acceptable sherry, giving him a double share.

"How singular! I see both a triple of mutual acquaintances and a triple of mutual strangers," said I, attempting clumsily to break the ice.

"A triple cannot be singular," Soames muttered, but at a gesture from me he held his peace. I topped up his glass.

Beatrix asked me for an explanation, and I hastened to comply. "You, Aubrey, and I each know the other two: a triple of mutual acquaintances."

"I think we are more than mere acquaintances, John," she replied.

"I am delighted to hear it, dear lady," said I, "but I was seeking a word that might apply to any pair of people. In contrast, Soames, you, and Dorinda are mutual strangers, in the sense that you have not met socially until now. Of course, Soames's fame has gone before him."

"It has indeed," said Crispin, giving me a sour look.

"Now, I find this fact somewhat remarkable—"

"You should not, Watsup," Soames interrupted. "At least, you should not consider it remarkable that at least one such triple, acquaintances or strangers, should occur."

"Why not?" asked Aubrey.

"Because at least one such triple must occur *whenever* six people are gathered together," Soames replied. "It matters not who knows whom."

"Well, I'll be dashed," said Aubrey. "Remarkable, what?"

"How can you be so sure, Mr Soames?" Beatrix enquired, her eyes shining—and not just, I suspected from the sherry.

"Because, my dear madam, it can be proved."

"Oh. Do go on, Mr Soames. I find such matters fascinating." Soames inclined his head, but I detected a faint, fleeting smile. He pretends to be immune to feminine charms, but I know this to be a sham. He merely lacks confidence. I hoped this would continue, for Beatrix is pretty and modest and would make a good catch for any compatible man. Myself, for example.

"The proof can be understood most simply by way of a diagram," said Soames. He rose, walked across to the dining table, and picked up some side plates and some cutlery, brushing aside my protestations along with several napkins, the mustard, and a potted aspidistra.

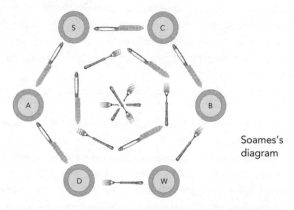

Soames's diagram

"The plates represent the six of us," Soames said, writing our initials on them with a stick of greasepaint, which he had presumably kept as a souvenir from the time he had contemplated a career on the stage. "A fork connecting two people denotes that they are acquaintances; a knife denotes they are strangers."

"Looking daggers at each other, then," said Beatrix. I hastened to applaud her wit and refill her glass.

"For instance, Watsup and I are joined by a fork at the centre of the table, but I am joined to everyone else by a knife.

"Now, as Watsup has so astutely pointed out, WAB is a triangle of forks and SBD is one of knives. My contention, however, is that *however* we arrange knives and forks, there will be at least one triangle formed by identical types of cutlery."

"Could it be both, Mr Soames?" asked Beatrix. Her eyes followed his every movement.

"Sometimes, madam, but not always. To go to extremes, if all implements are forks then there is no triangle of knives, and if they are all knives then there is no triangle of forks."

Beatrix nodded, a serious look on her face. "It seems, then," she mused, "that as forks are replaced by knives, and the opportunity to form a triangle of forks diminishes, that of forming a triangle of knives increases."

Soames nodded. "Very well put, madam. The proof is merely a matter of showing that the latter comes to pass before the former ceases. To be precise, let us choose one plate. Any plate. Five implements point towards it. At least three of them must be of the same kind. Why?"

"Because if there were two or fewer of each, there would be at most four implements," Beatrix said at once.

"Very good!" said I, before Soames could offer a similar compliment.

"Now," said he, "let us consider a set of three identical implements—let us assume they are forks, the knife case is similar—and look at the plates to which they point. Other than the original choice, of course. Now, either one of those plates is joined to another by a fork, or—"

"All three are joined by knives!" she cried. "In the first case, we have found a triangle of forks; in the second, one of knives. Why, Mr Soames, now that you explain it so clearly it seems—"

"Absurdly obvious," sighed Soames, taking a large sip of sherry.

This remark somewhat deflated her, and I waved my hand in apology for my friend's rudeness. Her instant smile was most gratifying.

This area of mathematics is called Ramsey Theory. See page 288 for further information.

How to Write Very Big Numbers

How many grains of sand are there in the universe? The greatest of the ancient Greek mathematicians, Archimedes, decided to combat the prevailing belief that the answer was infinite, by finding a way to express very large numbers. His book *The Sand Reckoner* assumed that the universe was the size that Greek philosophers thought it was, and that it was entirely filled with sand. He calculated that there would be (in our decimal notation) at most $1,000, \ldots ,000$ grains of sand, with 63 successive zeros.

That's big, but not infinite. Are there bigger numbers?

Mathematicians know that there is no largest (whole) number. They can be as big as you wish. The reason is simple: if there were a largest number, you could make it larger by adding 1. Most children who have mastered decimal notation quickly realise that you can always make a number bigger (indeed, ten times bigger) by putting an extra 0 on the end.

However, although there is no limit in principle to how large a number can be, there are often practical limitations in the way we choose to write numbers. For instance, the Romans wrote numbers using the letters I (1), V(5), X (10), L (50), C (100), D (500), and M (1,000), combining them in groups to get intermediate numbers. So 1-4 were written I, II, III, IIII, except that IIII was often replaced by IV (5 *minus* 1). In this system the largest number you can write is:

$$MMMMCMXCIX = 4,999$$

and you can knock a thousand off that if you stop with three M's.

However, the Romans sometimes needed bigger numbers. To symbolise a million, they put a bar (their name was *vinculum*) over the top of an M, to get $\overline{M}$. In general, a bar over a number multiplied its value by a thousand, but they seldom used this notation. When they did, they used only one bar, so a few million was as far as they could get. The limitations of their symbol system show that the size of numbers you can write down depends on the notation you use.

Nowadays we can go a lot further. A million is 1,000,000—pathetic. We can get *much* bigger numbers just by putting more zeros on the end, and moving the commas if necessary to keep the standard grouping into sets of three digits. (Mathematicians generally omit the commas anyway, but they often replace them by thin spaces: 1 000 000.) In the Western world, the standard dictionary names for large numbers reflect this practice: they start with million, billion, trillion, . . . and stop at centillion. Since human beings never manage to keep things simple, especially when it comes to mathematics, these words have (or at least used to have) different meanings on different sides of the Atlantic. In the USA a billion is 1,000,000,000, but in the UK it was 1,000,000,000,000—which Americans would call a trillion. But in today's interconnected world, the American usage has prevailed, perhaps because 'milliard'—the UK term for a thousand million—is (a) obsolescent and (b) too easily confused with 'million'. And a billion is a nice round number for international finance, at least until the world's banks threw away so much in the financial crisis that we had to get used to thinking in trillions.

A simpler way to write these numbers is to use powers of 10. So 10^6 denotes 1 followed by six 0's, which is a million. The 6 is called an *exponent*. A billion is 10^9 (or 10^{12} in old-fashioned UK-speak). A centillion turns out to be 10^{303} (10^{600} UK). Recognised extensions to the standard dictionary names go up to a millinillion, 10^{3003}. There are several systems and life is too short to go into them all, or indeed distinguish properly between them.

Two other names for large numbers, also found in most dictionaries, are *googol* and *googolplex*. A googol is 10^{100} (1 followed by a hundred 0s); the name was invented by James Newman's nine-year-old nephew Milton Sirotta. Sirotta also suggested a larger number, googolplex, which he defined as '1 followed by writing zeros until you get tired'. A certain lack of precision led to a refinement: '1 followed by a googol zeros'.

That's more interesting, because it runs into the same sort of problem that the Romans hit, except they got there a lot sooner. If you actually try to write a googolplex down in decimal notation, 1,000,000,000, . . . you won't get to the end during your lifetime. You wouldn't get to it during the lifetime of the entire universe to date. Assuming that conventional cosmological calculations are correct, you probably wouldn't get to it before the universe ended. In any case, you'd run out of space to write all those zeros, even if each were the size of a quark.

However, there is a compact way to write a googolplex: iterated exponents. Namely,

$$10^{10^{100}}$$

And once you start thinking along those lines, you can get to some very big numbers indeed. In 1976 computer scientist Donald Knuth invented a notation for very large numbers, which among other things turn up in some areas of theoretical computer science. When I say 'very large' I mean *very* large—so large that there's no way even to begin to write them down in conventional notation. The googolplex, which is 1 followed by 10^{100} zeros, is dwarfed by most of the numbers you can express using Knuth's *up-arrow* notation.

Knuth starts by writing

$$a \uparrow b = a^b$$

So, for example, $10\uparrow 2 = 100$, $10\uparrow 3 = 1000$, $10\uparrow 100$ is a googol, and $10\uparrow(10\uparrow 100)$ is a googolplex. The usual convention about the order in which exponents are taken (start at the right and work leftwards) lets us write this more simply as $10\uparrow 10\uparrow 100$. You don't need a lot of imagination to come up with $10\uparrow 10\uparrow 10\uparrow 10\uparrow 10\uparrow 10\uparrow 10$, say.

But that is just the start. Let

$$a\uparrow\uparrow b = a\uparrow\ a\uparrow\cdots\uparrow\ a$$

where a occurs b times. The exponentials are evaluated starting from the right, so that (for example)

$$a\uparrow\uparrow 4 = a\uparrow(a\uparrow(a\uparrow a))$$

For example,

$$2\uparrow\uparrow 4 = 2\uparrow(2\uparrow(2\uparrow 2)) = 2\uparrow(2\uparrow 4) = 2\uparrow 16 = 65{,}536$$

and

$$3\uparrow\uparrow 3 = 3\uparrow 3\uparrow 3 = 3\uparrow 27 = 7{,}625{,}597{,}484{,}987$$

The numbers rapidly become impossible to write down digit by digit. For instance, $4\uparrow\uparrow 4$ has 155 decimal digits. But that's the *point*: up-arrow notation provides a compact way to specify gigantic numbers. However, we've barely started. Now let

$$a\uparrow\uparrow\uparrow b = a\uparrow\uparrow\ a\uparrow\uparrow\cdots\uparrow\uparrow\ a$$

where a occurs b times on the right-hand side. Again the $\uparrow\uparrow$s are evaluated starting from the right. You get the picture: we can continue with

$$a\uparrow\uparrow\uparrow\uparrow b = a\uparrow\uparrow\uparrow\ a\uparrow\uparrow\uparrow\cdots\uparrow\uparrow\uparrow\ a$$
$$a\uparrow\uparrow\uparrow\uparrow\uparrow b = a\uparrow\uparrow\uparrow\uparrow\ a\uparrow\uparrow\uparrow\uparrow\cdots\uparrow\uparrow\uparrow\uparrow\ a$$

and so on, where as always a occurs b times and we evaluate starting from the right.

R.L. Goodstein developed Knuth's notation and simplified it, leading to expressions he called hyperoperators. John Conway developed a similar 'chained arrow' notation with horizontal arrows and brackets.

In string theory, an area of theoretical physics that aims to unify relativity with quantum mechanics, the number

10↑10↑500 turns up: it is the number of potentially different structures for space-time. According to Don Page, the longest finite time explicitly calculated by a physicist is a paltry

$$10{\uparrow}10{\uparrow}10{\uparrow}10{\uparrow}10{\uparrow}1.1 \text{ years}$$

This is the Poincaré recurrence time for the quantum state of a black hole with the mass of the entire universe; that is, the time it would take before this system returned to its original state and, in effect, history started to repeat itself.

* *

Graham's Number

Mathematicians occasionally need bigger numbers than physicists do. Not just for the fun of it: because these numbers actually turn up in sensible problems. Graham's number, named for the American Ron Graham, arises in combinatorics—the mathematics of counting different ways to arrange objects or fulfil conditions.

In 1978 Graham and Bruce Rothschild were working on a problem about hypercubes, multidimensional analogues of the cube. A square has 4 corners, a cube has 8, a four-dimensional hypercube has 16, and an n-dimensional hypercube has 2^n corners. They correspond to all possible sequences of n 0's and 1's in a system of n coordinates.

Take an n-dimensional hypercube, and draw lines connecting *all* pairs of corners. Colour each edge either red or blue. What is the smallest value of n for which *every* such colouring contains at least one set of four corners, lying in a plane, such that all the edges joining them have the same colour?

The two mathematicians proved that such a number n exists, which is far from obvious. Graham had earlier found a simpler proof, but he had to use a larger number: in Knuth's up-arrow notation it is at most:

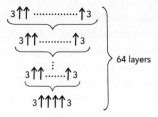

Here the numbers below the horizontal braces show how many arrows occur above the brace. Work backwards from the bottom layer: there are $3\uparrow\uparrow\uparrow\uparrow3$ up-arrows in the previous (63rd) layer. Then use *that* number of arrows in the 62nd layer to get a new number. Then use *that* number of arrows in the 61st layer...! It's not possible to write down any of these numbers in standard decimal notation, sorry. They are far worse than a googolplex in that respect. But that's their charm...

This is Graham's number, and it is truly gigantic. And then some. The value found by Graham and Rothschild is smaller, but still ridiculously large, and harder to explain, so I won't.

Ironically, workers in the area conjecture that the number can be made *much* smaller. In fact, that $n = 13$ will do. But this has not yet been proved. Graham and Rothschild proved that n must be at least 6; Geoff Exoo raised that to 11 in 2003; the best result now is that n must be at least 13, proved in 2008 by Jerome Barkley.

See page 289 for further information.

Can't Wrap My Head Around It

When scientists mention large numbers, such as the age of the universe (13.798 billion years or about 4.35 sextillion seconds) or the distance to the nearest star (0.237 light years or about 2.24 trillion kilometres) we all tend to say things like "I can't wrap my head around that". The same goes for the cost of the global financial crisis, one of the higher estimates being £1.162 trillion to the UK economy.* Let's say a round trillion, £10^{12}.

Millions, billions, trillions—in many people's minds, these are all pretty much the same: too big to wrap your head around.

This inability to internalise big numbers affects our views on many things, especially politics. There was much protest, especially from the airline industry, when Eyjafjallajökull in Iceland spewed out volcanic ash and grounded most of the UK's aeroplanes. (I wasn't too happy myself: I was due to fly to Edinburgh and had to change plans rapidly and drive instead.) The cost was estimated at £100 million per day: £10^8.

To be fair, that was the loss to a relatively small number of companies. But the scale of the outcry was probably greater than that caused by the financial crisis.

The big secret about comparing large numbers is that you don't *have* to wrap your head round them. Indeed, it is probably best not to. The mathematics—indeed, basic arithmetic—does it for you. For example, we can ask how long the flying ban would have to continue in order to cost the economy the same amount that the banking crisis did. The calculation goes:

> Cost of banking crisis: £10^{12}
> Cost per day of volcano: £10^8
> $10^{12}/10^8 = 10^4$ days $= 27$ years

I find this period of time eminently graspable, and I have no trouble recognising that it is a great deal longer than one day. So I can *work out* that the flying ban would have needed to go on for 27 years before it did as much economic damage as the bank crisis, without wrapping my head round the larger figures involved in the calculation.

This is what mathematics is for. Don't wrap your head round things: do the maths.

• •

* This figure was more than the final cost because the banks paid money back and some of it was temporary support. By March 2011 it was £450 billion, roughly half as big.

The Affair of the Above-Average Driver

From the Memoirs of Dr Watsup

I threw the newspaper on to the table in disgust. "I say, Soames—look at this ridiculous statistic!"

Hemlock Soames grunted, and concentrated on lighting his pipe.

"Seventy-five percent of hansom cab drivers think their ability is above average!"

Soames looked up. "What's ridiculous about that, Watsup?"

"Well, I—Soames, it's just not possible! They must have an exaggerated opinion of themselves!"

"Why?"

"Because the average has to be in the middle."

Hansom cab from John Thompson and Adolphe Smith, *Street Life in London*, 1877

The detective sighed. "A common misconception, Watsup."

"Miscon— what's wrong with it?"

"Just about everything, Watsup. Suppose 100 people are assessed on a score ranging from 0 to 10. If 99 of them score 10 and the other scores 0, what is the average?"

"Uh...990/100...which is 9.9, Soames."

"And how many are above average?"

"Uh...99 of them."

"As I said, a misconception."

I was not so readily diverted. "But the excess is small, Soames, and the data are not typical."

"I exaggerated the effect to demonstrate its existence, Watsup. Any data that are skewed—asymmetric—are likely to behave in a similar manner. For example, suppose that most drivers are reasonably competent, a significant minority is appallingly bad, and a very tiny number is excellent. Which drivers are above average in such circumstances?"

"Well...the bad ones bring the average down, and the excellent ones don't compensate...My word! The competent and excellent drivers are *all* above average!"

"Indeed," Soames replied. He rapidly sketched a graph on a sheet of scrap paper. "With these data, which are more realistic, the average is 6.25, and 60% of the drivers lie above that."

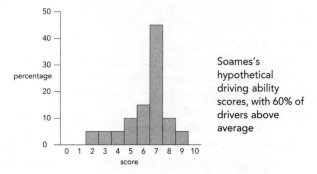

Soames's hypothetical driving ability scores, with 60% of drivers above average

"So the article in the *Manchester Mirrograph* is wrong?" I enquired.

"Are you surprised, Watsup? Very few of its articles are correct, to be frank. But this one falls into a common trap. It confuses the average with the median—which is *defined* to be the value for which half are above and half below. The two often differ."

"So it is not possible for 75% of drivers to be above the median?"

"Only if the number of drivers is zero."

"But 75% of drivers *could* be above average?"

"Yes."

"And they *wouldn't* have an unduly high opinion of their own abilities?"

Soames sighed again. "That, my dear Watsup, is a kettle of horses of a different colour of fish. There is a common form of cognitive bias called illusory superiority. People imagine themselves to be superior to others, even when they are not. Almost all of us suffer from this bias, with the notable exception of myself. An article in *Quantitative Phrenology and Cognition* last month reported that 69% of Swedish cab-drivers rated themselves above the *median*. And that is definitely illusory."

See page 289 for genuine modern data.

The Mousetrap Cube

Jeremiah Farrell invented a magic word *cube* obeying similar principles to those governing his magic word squares, see page 20. Here the word involved is MOUSETRAP and the magic numbering of the letters goes M = 0, O = 0, U = 2, S = 6, E = 9, T = 18, R = 3, A = 1, P = 0. Some of the words are personal names, and others are *very* obscure. For instance, OSE is the name of a demon—and also of places in Japan, Nigeria, Poland, Norway, and Skye. Still, the amazing thing is that it can be done at all.

top layer		
MOP	RUE	SAT
RAT	SOP	EMU
USE	MAT	PRO

middle layer		
EAR	SOT	UMP
SUP	MAE	ROT
TOM	PUR	SEA

bottom layer		
STU	MAP	ORE
MOE	RUT	SAP
RAP	OSE	TUM

Successive layers of the mousetrap magic word cube

Sierpiński Numbers

Number theorists seeking large primes often consider numbers of the form $k2^n + 1$, for specific choices of k, as n varies. Experiment suggests that for most choices of k, these numbers include a least one prime, often more. For instance, if $k = 1$ then $1 \times 2^n + 1$ is prime when $n = 2, 4, 8$. If $k = 3$ then $3 \times 2^n + 1$ is prime when $n = 1, 2, 5, 6, 8, 12$. If $k = 5$ then $5 \times 2^n + 1$ is prime when $n = 1, 3, 7$. (In general, we can divide k by any factors of 2 to make it odd and include these factors in the 2^n. So we may assume k is odd without losing generality. For example, $24 \times 2^n = 3 \times 2^3 \times 2^n = 3 \times 2^{n+3}$.)

It is tempting to conjecture that for any $k \geq 2$ there exists at least one prime of the form $k2^n + 1$. However, in 1960 Wacław Sierpiński proved that there exist infinitely many odd numbers k for which *all* numbers of the form $k2^n + 1$ are composite. These are called Sierpiński numbers.

In 1992 John Selfridge proved that 78,557 is a Sierpiński number, by showing that all numbers of the form $78,557 \times 2^n + 1$ are divisible by at least one of the numbers 3, 5, 7, 13, 19, 37, 73. These are said to form a covering set. The first ten known Sierpiński numbers are:

> 78,557   271,129   271,577   322,523   327,739
> 482,719   575,041   603,713   903,983   934,909

It is widely believed that 78,557 is the smallest Sierpiński number, but that has not yet been proved or disproved. Since 2002 the website www.seventeenorbust.com has been organising a search for primes of the form $k2^n + 1$, whose existence would prove that k is *not* a Sierpiński number. When the search started, there were 17 possible Sierpiński numbers smaller than 78,557, but these have been eliminated one by one until just six remain: 10,223, 21,181, 22,699, 24,737, 55,459, and 67,607. Along the way the project has discovered several very large primes.

k	eliminated by prime of the form $k2^n + 1$
4847	$4847 \times 2^{3321063} + 1$
5359	$5359 \times 2^{5054502} + 1$
	(at the time the fourth largest known prime)
10223	
19249	$19249 \times 2^{13018586} + 1$
21181	
22699	
24737	
27653	$27653 \times 2^{9167433} + 1$
28433	$28433 \times 2^{7830457} + 1$
33661	$33661 \times 2^{7031232} + 1$
44131	$44131 \times 2^{995972} + 1$
46157	$46157 \times 2^{698207} + 1$
54767	$54767 \times 2^{1337287} + 1$
55459	
65567	$65567 \times 2^{1013803} + 1$
67607	
69109	$69109 \times 2^{1157446} + 1$

James Joseph Who?

James Joseph Sylvester

James Joseph Sylvester was an English mathematician, who worked with Arthur Cayley in matrix theory and invariant theory, among other areas. He had a lifelong interest in poetry, and often put extracts from poems in his mathematical research articles. He moved to the USA in 1841, but returned shortly after. In 1877 he crossed the Atlantic again, and took up a position as the first Professor of Mathematics at Johns Hopkins University, and founded the *American Journal of Mathematics*, still going strong. He returned to England in 1883.

His name was originally James Joseph. When his elder brother emigrated to the USA, the immigration officials told him he had to have three names: first, middle, and surname. For some reason the brother added 'Sylvester' as a new surname. So James Joseph did the same.

. . . .

The Baffleham Burglary 🔍

From the Memoirs of Dr Watsup

Lord Baffleham's stately home had been burgled and some emeralds and rubies stolen from the safe. Soames, called in to investigate, quickly became suspicious of two visitors, Lady Esmeralda Nickett and Baroness Ruby Robham. Both had fallen on hard times, and had no doubt succumbed to temptation. But where was the proof?

Both ladies admitted possessing some jewels, but claimed them to be their own. Soames had not yet persuaded Inspector Roulade to obtain a search warrant, which might settle the matter, and had not been able to inspect the ladies' jewel boxes.

"The case," said Soames, "hinges on just how many jewels the two ladies possess. If the numbers match what was stolen, we have the final piece of evidence we need. Roulade is willing to request a search warrant, but only if we can tell him those two numbers."

"Esmeralda stated that she has only emeralds," I muttered, half to myself. "And Ruby says she has only rubies."

"Indeed. I am sure those statements are true. Now, the testimony of the maidservant places the number of each jewel somewhere between 2 and 101, not excluding those two numbers."

"The cook is reluctant to tell tales," I said. "But I have persuaded her to tell me the product of the two numbers."

"And the butler, equally reticent but open to persuasion in the form of ten gold sovereigns, has told me their sum," Soames replied.

"Then we can work out the two numbers by solving a quadratic equation!" I cried in excitement.

"Of course, though we won't know which number applied to the emeralds and which to the rubies," mused Soames. "The data are symmetric. But a match either way would be enough for Inspector Roulade to obtain a search warrant, which will no doubt suffice."

"If you tell me the product," I said, "I can solve the equation."

"Ah, my dear Watsup, you are *so* unsubtle," Soames complained. "Let me see if I can deduce the numbers without your telling me . . . Now, do you know what the two numbers are?"

"No."

"I knew that," said Soames, to my annoyance. If so, *why ask?* But suddenly light dawned.

"Now I know the numbers," I told him.

"In that case, so do I, Watsup."

What were the two numbers? See page 290 for the answer.

⸱ ⸱

The Quadrillionth Digit of π

We currently know the decimal expansion of π to 12,100,000,000,050 digits, a calculation performed by Shigeru Kondo in 2013 over a period of 94 days. No one really cares what the answer is, but this kind of record-breaking has led to some remarkable new insights, as well as being a good way to test new supercomputers. One of the more curious discoveries is that it is possible to compute specific digits of π without finding the previous ones. But for now we can do this only in base-16, or hexadecimal, notation from which digits in bases 8, 4, and 2 (binary) can immediately be deduced. This idea generalises to constants other than π, and to base 3, but there is no systematic theory yet. Nothing like it is known for decimal notation, base 10.

The initial discovery, the BBP formula, is stated below: see also *Cabinet* page 210. It is an infinite series that makes it possible to calculate a specific hexadecimal digit of π *without* calculating any of the previous ones. So we can be confident that the quadrillionth binary digit of π is 0, thanks to Project PiHex, and going even further the two-quadrillionth binary digit of π is also 0, thanks to a 23-day long computation by one of Yahoo!'s employees. Despite that, it would take another equally massive computation to find the previous digit.

In 2011 David Bailey, Jonathan Borwein, Andrew Mattingly, and Glenn Wightwick wrote a survey of this area [The computation of previously inaccessible digits of π^2 and Catalan's constant, *Notices of the American Mathematical Society* 60 (2013) 844–854]. They described how to compute base-64 digits of π^2, base-729 digits of π^2, and base-4096 digits of a number called Catalan's constant, starting at the 10 trillionth place.

The story starts with a series known to Euler:

$$\log 2 = \frac{1}{2} + \frac{1}{2.4} + \frac{1}{3.8} + \frac{1}{5.16} + \cdots = \sum_{k=1}^{\infty} \frac{1}{k2^k}$$

in Σ-notation for sums. Because of the powers of 2 that occur, this series can be converted into a method for computing specific binary digits of log 2. The computations remain feasible, but take much longer, as the position of the digit concerned gets larger.

The BBP (Bailey-Borwein-Plouffe) formula is

$$\pi = \sum_{n=0}^{\infty} \left(\frac{4}{8n+1} - \frac{2}{8n+4} - \frac{1}{8n+5} - \frac{1}{8n+6} \right) \left(\frac{1}{16} \right)^n$$

and the powers of 16 make it possible to compute specific hexadecimal digits of π. Since $16 = 2^4$, the series can also be used to find binary digits.

Is this approach limited to these two constants? From 1997 onwards, mathematicians sought similar infinite series for other constants, and succeeded for a large number of them, including

$$\pi^2 \quad \log^2 2 \quad \pi \log 2 \quad \zeta(3) \quad \pi^3 \quad \log^3 2 \quad \pi^2\log 2 \quad \pi^4 \quad \zeta(5)$$

where

$$\zeta(n) = \frac{1}{1^n} + \frac{1}{2^n} + \frac{1}{3^n} + \frac{1}{4^n} + \frac{1}{5^n} + \cdots$$

is the Riemann zeta function. They also succeeded for Catalan's constant

$$G = \frac{1}{1^2} - \frac{1}{3^2} + \frac{1}{5^2} - \frac{1}{7^2} + \frac{1}{9^2} + \cdots$$
$$= 0.9115965599417722\ldots$$

Some of these series yield digits to base 3 or some power of 3. For example, the amazing formula of David Broadhurst

$$\pi^2 = \frac{2}{27} \sum_{k=0}^{\infty} \frac{1}{729^k} \left(\frac{243}{(12k+1)^2} - \frac{405}{(12k+2)^2} - \frac{81}{(12k+4)^2} \right.$$
$$- \frac{27}{(12k+5)^2} - \frac{72}{(12k+6)^2} - \frac{9}{(12k+7)^2} - \frac{9}{(12k+8)^2}$$
$$\left. - \frac{5}{(12k+10)^2} + \frac{1}{(12k+11)^2} \right)$$

can be used to compute digits of π^2 to base $729 = 3^6$.

Is π Normal?

The digits of π look random, but they can't be truly random since you always get the same numbers every time you calculate π, barring errors. It is generally thought that, like almost all random sequences of digits, *every* finite sequence occurs somewhere in the decimal expansion of π. Indeed, infinitely often, though with lots of junk in between successive occurrences, and in the same proportion you'd expect for a random sequence.

It can be proved that this property, called normality, holds for 'almost all' numbers: in any sufficiently large range of numbers, the proportion that are normal gets as close as we wish to 100%. But this leaves a loophole, because any given number, in particular π, might be an exception. Is it? We don't know. Until recently the question looked hopeless, but formulas like the ones above have opened up a new line of attack, which might just solve the question for binary (or hexadecimal) digits.

The link arises through another mathematical procedure: iteration. Here we start with a number, apply some rule to get another one, and repeatedly apply that rule to get a sequence of numbers. For example, if we start with 2 and the rule is 'form the square' the sequence goes

$$2 \quad 4 \quad 16 \quad 256 \quad 65,636 \quad 4,294,967,296 \quad \ldots$$

The binary digits of a number like log 2 can be generated by the iterative formula

$$x_{n+1} = 2x_n + \frac{1}{n} \quad (\text{mod } 1)$$

starting from $x_0 = 0$. The symbols (mod 1) mean 'subtract the integer part', so $\pi \,(\text{mod } 1) = 0.14159\ldots$. This formula would lead to a proof that log 2 is normal to base 2 if it could be shown that the resulting numbers are uniformly spread over the range from 0 to 1. Such 'equidistribution' is quite common. Unfortunately no one knows how to prove it holds for the iterative formula above, but it's a promising idea and might get there eventually.

There is a similar but more complicated iterative formula for π:

$$x_{n+1} = \left(16x_n + \frac{120n^2 - 89n + 16}{512n^4 - 1024n^3 + 712n^2 - 206n + 21}\right) \quad (\text{mod } 1)$$

If this is equidistributed, π is normal in binary.

This leads to a final, very strange, discovery. Suppose we stretch out the range from 0 to 1 by a factor of 16, so that

$y_n = 16x_n$ runs from 0 to 16. Then the integer parts of successive y_n range from 0 to 15. Experimentally, these numbers are *precisely* the successive hexadecimal digits of $\pi - 3$. This has been checked on a computer for the first 10 million places. In effect, this appears to provide a formula for the nth hexadecimal digit of π. The computation gets harder and harder the further you go, and took 120 hours.

There are solid reasons to expect this statement to be true, but they don't amount to a rigorous proof. It is known that very few errors, if any, occur. Since none occurs for the first 10 million iterations, there is about a one in a billion chance that a later error will occur. However, that's not a proof—just an excellent reason to hope that one can be found.

A final conjecture, also based on sound evidence, shows how strange this area is. Namely: nothing like this can be done for the other well-known constant e, the base of natural logarithms, roughly equal to 2.71828. There seems to be something special about π compared to e.

* *

A Mathematician, a Statistician, and an Engineer...

...went to the races. Afterwards, they met in the bar. The engineer was drowning his sorrows. "I can't understand how I lost all my money. I measured the horses, calculated which was mechanically most efficient and robust, and figured out how fast they could run—"

"That's all very well," said the statistician, "but you forgot individual variability. I did a statistical analysis of their previous races, and used Bayesian methods and maximum likelihood estimators to find which horse had the greatest probability of winning."

"And did it?"

"No."

"Let me buy you guys a drink," said the mathematician, pulling out a bulging wallet. "I did pretty well today."

Obviously here was a man who knows something about horses. The others insisted on being told his secret.

Reluctantly he agreed. "Consider an infinite number of identical spherical horses . . . "

* *

Lakes of Wada

Topology is often counterintuitive. This makes it difficult, but also interesting. Here's a strange topological fact with applications to numerical analysis.

Two regions of the plane can share a common boundary curve; think of the English–Scottish border or the American–Canadian one. Three or more regions can share a common boundary *point*: at Four Corners in America, the states Arizona, Colorado, New Mexico, and Utah all meet.

Four Corners

With some ingenuity any number of regions can be arranged to share *two* common boundary points. But it doesn't seem possible for three or more regions to have more than two boundary points in common. Let alone for them all to have exactly the same boundary.

However, it can be done.

First, we have to be precise about what a boundary point is. Suppose we have some region in the plane. It need not be a polygon: it can have a very complicated shape—any collection of points whatsoever. Say that a point lies in the *closure* of the region if *every* circular disc with centre at that point and nonzero radius (however small) contains some point in the region. Say that a point lies in the *interior* of the region if some circular disc with centre at that point and nonzero radius is contained in the region. Now the *boundary* of the region consists of all points in its closure that do not lie in its interior.

Got that? Stuff on the edge but not inside, basically.

For a polygonal region, bounded by a series of straight line segments, the boundary consists of those line segments, so what we have defined agrees with the usual concept in this case. It can be proved that three or more polygonal regions cannot all have the same boundary. But this is not true for more complicated regions. In 1917 the Japanese mathematician Kunizō Yoneyama published an example of three regions that have *exactly the same boundary*. He said that his teacher Takeo Wada had come up with the idea. Accordingly, the regions (or anything of the same kind) are known as the Lakes of Wada.

We construct the three regions step by step using an infinite process. Begin with three square regions.

Start with three squares...

Then extend the first region by adding a trench that wraps round all three regions. Do this so that every point on the boundary of any square lies close to the trench. Also, make sure that the trench does not close up on itself to leave a hole in the resulting region.

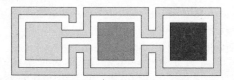

Dig a trench...

Then extend the second region by adding a yet narrower trench that wraps round all three regions constructed so far.

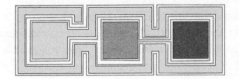

Dig a thinner trench...

Continue like that, with an even narrower trench from the third region. Then go back to the first region adding a still narrower trench, and so on.

Repeat this construction infinitely often. The resulting regions have infinitely complicated, infinitely thin trenches. But because each successive region gets closer and closer to *everything* previously constructed, all three regions have the same (infinitely complicated) boundary.

The same idea works if we start with four regions or more: *all* of the regions constructed have the same boundary.

The Lakes of Wada were originally invented to show that the topology of the plane is not as straightforward as we might imagine. Many years later, it was discovered that such regions arise naturally in numerical methods for solving algebraic equations. The cubic equation $x^3 = 1$, for example, has only one real solution $x = 1$, but it also has two *complex* solutions $x = -\frac{1}{2} + \frac{1}{2}i\sqrt{3}$ and $x = -\frac{1}{2} - \frac{1}{2}i\sqrt{3}$, where $i = \sqrt{-1}$. The complex numbers can be visualised as points in a plane, with $x + iy$ corresponding to the point with coordinates (x,y).

A standard method for finding numerical approximations to a solution begins with a randomly chosen complex number, then calculates a second number in a specific manner, and repeats until the numbers get very close together. The result is then close to a solution. Which of the three solutions it approaches depends on where you start, and it does so in a very intricate way. Suppose we colour points in the complex plane according to which solution they lead to: say medium grey if the solution is $x = 1$, light grey if the solution is $x = -\frac{1}{2} + \frac{1}{2}i\sqrt{3}$, and dark grey if the solution is $x = -\frac{1}{2} - \frac{1}{2}i\sqrt{3}$. Then the points that are coloured a given shade of grey define a region, and it can be proved that all three regions have *the same boundary*.

Unlike Wada's construction, the regions here are not connected: they break up into infinitely many separate pieces. However, it is striking that regions of such complexity arise naturally in such a basic problem of numerical analysis.

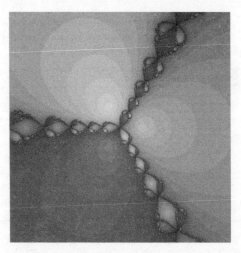

The three regions corresponding to solutions of the cubic equation

Fermat's Last Limerick

A challenge for many long ages
Had baffled the savants and sages.
Yet at last came the light:
Seems that Fermat was right—
To the margin add two hundred pages.

Malfatti's Mistake

From the Memoirs of Dr Watsup

"Extraordinary!" I exclaimed.

Soames glanced in my direction, obviously annoyed at being interrupted in his perusal of his extensive collection of plaster casts of squirrel footprints.

"The answer seems obvious—yet, apparently, it is wrong!" I cried.

"The obvious usually is," said Soames. "Wrong," he added by way of clarification.

"Ever heard of Gian Francesco Malfatti?" I asked.

"The multiple axe-murderer?"

"No, Soames, that was 'Hacker' Frank Macavity."

"Ah. My apologies, Watsup, you are of course correct. I am distracted. My specimen of *Ratufa macroura* tracks is disintegrating. The grizzled giant squirrel."

"Malfatti was an Italian geometer, Soames. In 1803 he asked how to cut three cylindrical columns from a wedge of marble in such a manner as to maximise their total volume. He assumed that the problem is equivalent to drawing three circles inside the triangular cross section of the wedge in such a manner as to maximise their total area."

"A naive but possibly correct assumption," Soames replied. "Though the columns might be cut on a slant."

"Oh, I had not... But let us suppose his assumption to be correct, since the question can always be suitably rephrased. It then seemed obvious to Malfatti that the solution must comprise three circles, each tangent to the other two and to two sides of the triangle." I drew a quick sketch.

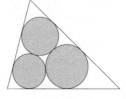

Malfatti circles

"I see the fallacy," said Soames, in that annoyingly offhand way that he often adopts when he has instantly grasped complexities beyond most other mortals.

"I confess I do not," I said. "For if a circle lies inside the triangle, does not overlap the others, and is *not* tangent in that manner, it can be enlarged."

"Correct," said Soames. "But that merely proves the sufficiency, not the necessity, of the tangency conditions."

"I am aware of that, Soames. But—how else might the circles be arranged?"

"There might be other ways for the tangency to occur, of course. For example, Watsup: have you considered the simplest case, that of an equilateral triangle?"

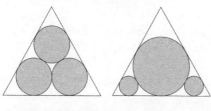

Two possible arrangements for an equilateral triangle

"First," said Soames, "there is Malfatti's arrangement, the left-hand figure. But what about the right-hand figure? Again, no circle can be enlarged, but the pattern of tangencies is different. The small circles are tangent to the large one, but not to each other. Instead, one cycle is tangent to three sides of the triangle."

I stared at the figures. "To the eye, Soames, the first arrangement has larger area."

He laughed. "Which only goes to show, Watsup, how easily deceived the eye is. Suppose that the triangle has sides of unit length. Then Malfatti's arrangement has area 0.31567, but the other one has area 0.31997, which is very slightly larger."

There are times when Soames's erudition leaves one breathless. "The difference may be small, Soames, but the implication is decisive. Malfatti was wrong."

"Indeed. Moreover, Watsup, the difference between Malfatti's arrangement and the correct one can sometimes be much greater. For example, if the triangle is long, thin, and isosceles, then the correct solution stacks the three circles on top of each other, and the area is almost double that of Malfatti's arrangement."

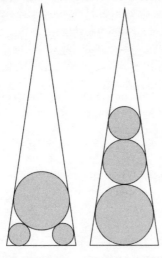

A thin isosceles triangle. *Left*: Malfatti's arrangement. *Right*: Largest area.

He paused to hurl the crumbled cast of the tracks of *Ratufa macroura* across the room into the fireplace. "The irony is," he added, "that Malfatti's arrangement is *never* the best. The greedy algorithm—fit the largest circle possible inside the triangle, then find the largest that fits into a remaining gap, and finally do the same for the third circle—is always superior, and indeed finds the correct answer."

See page 291 for further information.

See page 291 for further information.

● ●

Square Leftovers

Perfect squares end in one of the digits 0, 1, 4, 5, 6, or 9. They don't end in 2, 3, 7, or 8. In fact, the final digit of the square of a number depends only on the final digit of the number:

If a number ends in 0 then its square ends in 0.
If a number ends in 1 or 9 then its square ends in 1.
If a number ends in 2 or 8 then its square ends in 4.
If a number ends in 5 then its square ends in 5.
If a number ends in 4 or 6 then its square ends in 6.
If a number ends in 3 or 7 then its square ends in 9.

Number theorists prefer to phrase this kind of effect in terms of integers to some modulus (page 188). If the modulus is 10, then the only numbers that need to be considered are 0–9: the possible remainders on dividing any number by 10. Their squares (mod 10) are

0 1 4 9 6 5 6 9 4 1

and the list of rules for final digits of squares is a different way to say the same thing.

Aside from the initial 0, the list of squares (mod 10) is symmetric: the numbers 1 4 9 6 appear after 5 in reverse, 6 9 4 1. The symmetry arises because n and $10-n$ have the same squares modulo 10. Indeed, $10-n$ is the same as $-n$ (mod 10), and $n^2 = (-n)^2$. These four numbers therefore occur *twice* in the list; 0 and 5 occur once; but 2, 3, 7, 8 don't occur at all. It's not very democratic, but there you go.

What happens if we use a different modulus? The values of the squares, to that modulus, are called *quadratic residues*. (Here 'residue' refers to the remainder on dividing by the modulus.) The rest are *quadratic non-residues*.

For instance, suppose that the modulus is 11. Then the possible perfect squares (of the numbers less than 11) are

0 1 4 9 16 25 36 49 64 81 100

which, when reduced (mod 11), yield

0 1 4 9 5 3 3 5 9 4 1

So the quadratic residues (mod 11) are

0 1 3 4 5 9

and the non-residues are

2 6 7 8

Here's a short table:

modulus m	squares (mod m)	quadratic residues
2	0 1	0 1
3	0 1 1	0 1
4	0 1 0 1	0 1
5	0 1 4 4 1	0 1 4
6	0 1 4 3 4 1	0 1 3 4
7	0 1 4 2 2 4 1	0 1 2 4
8	0 1 4 1 0 1 4 1	0 1 4
9	0 1 4 0 7 7 0 4 1	0 1 4 7
10	0 1 4 9 6 5 6 9 4 1	0 1 4 5 6 9
11	0 1 4 9 5 3 3 5 9 4 1	0 1 3 4 5 9
12	0 1 4 9 4 1 0 1 4 9 4 1	0 1 4 9

At first sight, there are few clear patterns aside from those already mentioned. This is part of the charm of the area, in fact: while there are patterns, it takes some digging to find them. Several of the greatest mathematicians devoted a lot of attention to this area, among them Euler and Carl Friedrich Gauss.

When we square a number we multiply it by itself, and when it comes to multiplication, what matter most in number theory are the primes. So it pays to start with prime moduli: 2, 3, 5, 7, 11 in the above list. The modulus 2 is exceptional: the only possible residues modulo 2 are 0 and 1, and both are squares. For all other primes, about half of the residues are squares and the rest are not. More precisely, if p is prime then there are $(p+1)/2$ distinct quadratic residues and $(p-1)/2$ non-residues. The quadratic residues are usually squares of two distinct numbers, n^2 and $(-n)^2$ for suitable n. However, 0 occurs only once because $-0 = 0$.

Composite moduli complicate the story. Now the same quadratic residue can sometimes be the square of more than two numbers. For example 1 occurs four times for modulus 8, as the square of 1, 3, 5, and 7. The best way to make sense of all this is to use modern abstract algebra, but it's worth taking a look at modulus 15. This has two prime factors: $15 = 3 \times 5$. Now the list of squares is:

n	0	1	2	3	4	5	6	7	8	9	10	11	12	13	14
n^2	0	1	4	9	1	10	6	4	4	6	10	1	9	4	1

So the quadratic residues modulo 15 are

$$0 = 0^2$$
$$1 = 1^2, 4^2, 11^2, 14^2$$
$$4 = 2^2, 7^2, 8^2, 13^2$$
$$6 = 6^2, 9^2$$
$$9 = 3^2, 12^2$$
$$10 = 5^2, 10^2$$

Some residues occur once, some twice, some four times. Those that occur fewer than four times are squares of numbers that are divisible by either 3 or 5, the prime factors of 15. All the other numbers occur in sets of four, all having the same square.

This is a general pattern for any modulus of the form pq where p and q are distinct *odd* primes. The numbers between 0 and $pq-1$ that are not divisible by p or q split into sets of four, each set having the same square. (This fails if one of the primes is 2: for example $10 = 2 \times 5$ and we've already seen that in this case squares either occur in pairs or on their own.)

In algebra we get used to the idea that every positive number has *two* square roots: one positive, the other negative. But in arithmetic modulo pq, most numbers (those not divisible by p or q) have *four* distinct square roots.

This curious fact has a remarkable application, to which we now turn.

· ·

Coin Tossing over the Phone

Suppose that Alice and Bob want to play a coin-tossing game with a 50–50 outcome. As we've already seen (page 131) Alice is in Alice Springs and Bob is in Bobbington. Can they toss the coin over the phone? The big snag is the same as for poker. If Alice tosses a coin, or, equivalently, carries out any activity with two equally probable results, and tells the result to Bob, he has no idea whether she is telling the truth. Nowadays they could do it over Skype and watch the coin being tossed, but even then, the toss could be rigged by filming several tosses ahead of time and sending one of those instead.

Tossing a coin is like playing poker with a pack of just two cards, so they could adapt the method on page 132. However, there's another elegant way to achieve the same result, using quadratic residues. Here's how.

Alice chooses two large odd prime numbers p and q. She keeps them private, but sends their product $n = pq$ to Bob. You might think that Bob could then find p and q by factorising n, but as far as is currently known, there is no practical method to do that when the numbers are sufficiently large—say 100 digits for each of p and q. The fastest computer using the fastest known algorithm would take longer than the lifetime of the universe. So Bob must remain ignorant of the actual primes involved. However, there are very quick ways to test a 100-digit number to see whether it is prime. So Alice can find p and q by trial-and-error.

Bob chooses a random integer x (mod n), which he keeps private.

If he is extremely pedantic, he can quickly check whether x is a multiple of p and q: not by dividing by those numbers, since he doesn't know them, but by finding the hcf of x and n using Euclid's algorithm (page 110). If the result is not 1 he then knows either p or q, so the process has to be repeated with a new x. But in practice he need not bother, since when p and q have 100 digits the probability that p or q divides a randomly chosen x is 2×10^{-100}.

Bob now calculates x^2 (mod n), which can also be done quickly, and sends it to Alice. They have agreed that if Alice can correctly deduce either x or $-x$ she wins ('heads'). Otherwise, she loses ('tails').

By the previous item Alice knows that integers modulo pq that are not divisible by p or q have exactly four square roots.

Since x and $-x$ have the same square, these are of the form a, $-a$, b, $-b$ for suitable a and b. Alice knows p, q, and x^2, which implies that she can compute these four square roots quickly. Two of them must be Bob's x and $-x$; the other two must be different. So Alice has a 50% chance of guessing $\pm x$ correctly—equivalent to tossing a fair coin. She chooses one of these four, say b, and sends it to Bob.

Bob tells Alice whether $b = \pm x$ or not; that is, whether she is right or wrong.

Ah—but how do we stop Bob cheating? And how does Bob *know* that Alice has done what she is supposed to do?

Whether or not $b = \pm x$, Bob can be happy that Alice has played fairly by computing b^2 (mod n). This should be the same as x^2.

If Alice loses, she can convince herself that Bob has not lied by asking him to send her the prime factors p and q of n. Normally this would be impossible, but *if Alice has lost*, then Bob knows *all four* square roots of x^2, and there is a number-theoretic trick to calculate p and q quickly from this information. In fact, the highest common factor of $a + b$ and n is one of the two primes, and again it can be found using Euclid's algorithm. The other can then be found by division.

. . . .

How to Stop Unwanted Echoes

Quadratic residues may seem typical of the abstruse explorations of pure mathematicians: an intellectual game with no practical uses. But it's a mistake to think that a mathematical idea is useless just because it doesn't obviously derive from a practical problem in everyday life. It's also a mistake to think that everyday life is as straightforward as it appears to be on the surface. Even something as simple as a pot of jam in a supermarket involves making the glass, growing the sugar cane or beet, refining the sugar . . . and pretty soon you're into statistical hypothesis testing for disease-resistant fruit and the design of the ship used to transport various components or the finished product round the globe. In a world of 7 billion people, mass food production is not just a matter of picking some blackberries and boiling them up.

It's true that the mathematicians who first came up with these ideas had no particular applications in mind; they just thought quadratic residues were interesting. But they were also convinced that understanding them would be a powerful addition to the mathematical toolkit. Practical people can't use a tool unless it exists. And while it might seem to make sense to wait for an application before inventing a suitable tool, we'd still be sitting in caves if we'd done it that way. "Why are you wasting time bashing those rocks together, Ug? You should be bashing mammoths over the head with a stick like the other boys."

Quadratic residues have many different uses. One of my favourites is the design of concert halls. When music reflects off a flat ceiling, the result is a distinct echo, which distorts the sound and is generally unpleasant. On the other hand, a ceiling that absorbs sound makes the performance sound dead and fuzzy. To get good acoustics, the sound has to be allowed to bounce back, but as a diffuse spread of sounds rather than a sharp echo. So architects fit diffusers on the ceiling. The question is: what shape should the diffusers be?

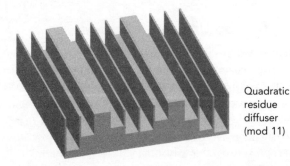

Quadratic residue diffuser (mod 11)

In 1975 Manfred Schroeder invented a diffuser consisting of a series of parallel grooves, whose depths are derived from the sequence of quadratic residues for some prime modulus. For example, suppose that the prime is 11. We've just seen that the squares of 0–10, reduced modulo 11, are:

    0 1 4 9 5 3 3 5 9 4 1

and the sequence repeats these values periodically for larger numbers. It's symmetric about the middle, between the two 3's, because $x^2 = (-x)^2$ modulo any prime. Compare the picture below, showing these numbers as rectangles, with the shape of the diffuser above. Notice that in this case the depths of the grooves are obtained by *subtracting* the residues from some constant depth. This has no serious effect on the main mathematical point.

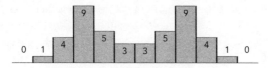

Graph of quadratic residues (mod 11)

What's so special about quadratic residues? One feature of a sound wave is its frequency: how many waves hit the ear every second. High frequencies give high notes, low frequencies give low notes. A related feature is the wavelength: the distance between successive peaks. High-frequency waves have short wavelength, and low-frequency waves have longer wavelength. Waves of a given wavelength tend to resonate with cavities in the surface whose size is similar to that wavelength. So waves with different frequencies react differently when they hit a surface.

The quadratic residue diffuser has a delightful mathematical property: waves with many different frequencies react to it in the same manner. Technically, its Fourier transform is constant across a range of frequencies. Schroeder pointed out an important consequence: this shape diffuses sound waves of many different frequencies in the same manner. In practice the widths of the grooves are chosen to avoid the range of wavelengths that humans can hear, and their depths are a specific multiple of the sequence of quadratic residues, related to the width.

When the grooves are parallel, as in the picture, the sound is diffused sideways, at right angles to the direction of the grooves. There is a two-dimensional analogue, a square array of rods also based on quadratic residues, and this diffuses the sound equally in all directions. Diffusers like that are often found in recording studios, to improve the sound balance and get rid of extraneous noises.

So although Euler and Gauss had no idea what their invention would be used for, or indeed whether it would ever be used for anything, it often plays a crucial role, behind the scenes, when you listen to recorded music—be it classical, jazz, country, rock, hip hop, crossover thrash metal, or whatever else takes your fancy.

See page 292 for further information.

• •

The Enigma of the Versatile Tile

From the Memoirs of Dr Watsup

"Solving a crime is often likened to fitting together the pieces of a jigsaw puzzle," Soames remarked out of the blue. The blue being a cloud of smoke that enveloped his head, emanating from his pipe.

"An apt simile!" said I, raising my head from my newspaper.

He smiled slyly. "Not so, Watsup. On the contrary, a very poor one. When investigating a crime, we do not know what the pieces might be, nor whether we are in possession of them all. Not knowing the puzzle, how can we be certain of the answer?"

"Surely, Soames, that becomes evident when enough of the known pieces fit together into an elegant pattern."

He sighed. "Ah, but there can be so many pieces, Watsup. And so many patterns. Deciding which is correct takes a certain... *je ne sais quoi*. But I don't know what."

At that moment there was a knock on the door and a woman rushed in.

"Beatrix!" I cried.

"Oh, John! It has been stolen!" And she rushed into my arms, sobbing. I did my best to comfort her, though I confess my own heart was racing.

After a time, she became calm. "Please help me, Mr Soames! It is a ruby pendant inherited from my late mother. I looked for it this morning, and it had gone!"

"Do not distress yourself, my dear," said I, patting her shoulder. "Soames and I will apprehend the thief and recover the jewel."

"Did you come by cab?" Soames asked.

"Yes. It is waiting outside."

"Then we shall lose no time in investigating the scene of the crime."

After half an hour crawling on the floor, sampling dust from the corners of several rooms, and inspecting the doorstep and flowerbeds, Soames shook his head. "No sign of a break-in, Miss Sheepshear. But there are small scratch marks on your jewel case. Very fresh, and not yours, for whoever made them is left-handed." He put the case down. "Have any strangers visited the house recently? Tradesmen, perhaps?"

"No . . . Oh! The tilers!"

Two men claiming to be tilers had come to the back door offering to renovate the bathroom. "It is the new fashion, Mr Soames. Plain white square tiles, into which has been cut a blue motif made from tiles of a more elaborate shape. The Dimworthys had theirs done last month, and father—" Her voice failed, overcome, close to tears. I took her hand.

"Are you in the habit of engaging unknown tradesmen at the doorstep?" Soames enquired.

"Why, no, Mr Soames. Ordinarily we would deal only with a reputable firm. But they are all booked solid for months. And these seemed honest, decent men."

"They always do. Did you leave either of them unattended?"

She thought for a moment. "Yes. The assistant was left to make measurements in the bathroom while his master showed me sample motifs."

"Ample time to steal a small but valuable item. They are clever: by not being greedy they ensured that the theft was not immediately noticed. Did they leave any documentation?"

"No."

"Have they returned since?"

"No, I am awaiting a written estimate for the work."

"I venture to predict that it will not arrive, madam. It is a *modus operandi* that in the trade we call a 'distraction burglary'."

Over the following week, a steady stream of ladies engaged Soames's services with similar tales. The tradesmen varied in appearance, but Soames was unsurprised. "Disguised," he said.

The breakthrough came on the thirteenth case, at the home of Mrs Amelia Fotherwell. Soames noticed a lump of mud adhering to the bathroom door, in which was embedded a small piece of bone. The composition of the mud and the nature of the bone led to a dirty backyard next to a sardine cannery in one of a maze of tiny streets behind the Albert Dock.

"So now we break in and search for evidence?" said I, reaching for my revolver.

"No: that might alert the thief. We return to Baker Street and assemble our case."

"Tell me, Watsup," he said, as we shared a bottle of port. "What features do all these thefts have in common?" I pointed out those that sprang to mind. "Very good. But you have omitted the most significant feature. The motifs. You have a list of them, no doubt?"

I took out my notebook. It read:

- *Mrs Wotton*: three tiles forming an equilateral triangle.
- *Beatrix*: four tiles forming a square.
- *Miss Makepiece*: four tiles forming a square with a square hole.
- *The Cranford twins*: four tiles forming a rectangle with a rectangular hole.

- *Mrs Broadside*: four tiles forming a convex hexagon.
- *Mrs Probert*: four tiles forming a convex pentagon.
- *Lady Cunningham*: four tiles forming an isosceles trapezium.
- *Miss Wilberforce*: four tiles forming a parallelogram.
- *Mrs McAndrew*: four tiles forming windmill sails.
- *Mrs Tushingham*: six tiles forming a hexagon with a hexagonal hole.
- *Miss Brown*: six tiles forming an equilateral triangle with triangles cut off the corners.
- *Dame Jenkin-Glazeworthy:* twelve tiles forming a dodecagon (regular 12-sided polygon) with a regular 12-sided star hole.
- *Mrs Fotherwell*: twelve tiles forming a dodecagon with 12-sided star hole shaped like the blade of a circular saw.

"A remarkable collection," said Soames. "I think it is time to send a Baker Street Irreducible to Inspector Roulade, asking him to raid the premises near Albert Dock."

"What are you expecting the police to find?"

"Recall, Watsup: each lady told us that her motif was made from a number of identical tiles."

"Yes."

"But the motifs are very varied, suggesting that although each motif used a single shape, different motifs required different shapes of tile. The ladies cannot describe the shape except as 'irregular', so we have no evidence that the same shape was employed for each motif. I therefore expect the police to find thirteen boxes of strangely shaped tiles: one for each motif."

After a couple of hours Mrs Soapsuds appeared. "Inspector Roulade, Mr Soames."

The Inspector entered, accompanied by a constable bearing a box. "I have placed two suspects under arrest," said the Inspector.

"Roland 'the rat' Ratzenberg and 'Bugface' McGinty."

"Yes, but how on Earth—oh, never mind. I can hold them for twenty-four hours. But the evidence is weak."

Soames looks shocked. "Surely you found all those boxes of tiles? Is not that one merely a sample?"

The Inspector shook his head. "No: it is their entirety."

Soames walked over to the box and opened it. It contained twelve identical tiles. "Oh," said he.

"It seems the case has collapsed," I ventured. "I cannot believe that such a varied list of motifs can all be made from a single shape of tile."

But Soames suddenly became more animated. "You may well be right," he said. "Unless..." He produced ruler and protractor and began measuring one of the tiles.

After a few moments, a smile stole across his features. "Clever!" said he. "*Very* clever." He turned towards me. "I have been extremely foolish, Watsup, and *assumed* when I should have maintained an open mind. Do you remember our topic of conversation just before Beatrix arrived in distress?"

"Um—jigsaw puzzles."

"Indeed. And this case hinges upon one of the most remarkable jigsaw puzzles I have ever encountered. Look at this tile."

"It seems a very ordinary quadrilateral," I said.

"No, Watsup: it is a very extraordinary quadrilateral. Allow me to demonstrate." And he drew a diagram.

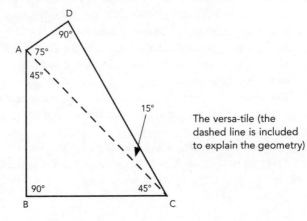

The versa-tile (the dashed line is included to explain the geometry)

"The sides AB and BC are equal, and ABC is a right angle, so angles BAC and BCA are 45°," Soames explained. "Angle ACD is 15°, so BCD is 60°. Angle ADC is again a right angle, and that makes angle CAD equal to 75°."

The Inspector and I remained unenlightened. Soames handed me four tiles. "Watsup, try fitting these together to make an elegant shape. Much as a detective might fit clues together to create an elegant deduction, to quote your earlier analogy."

"May I turn them over?"

"An excellent question! Yes, you may turn any piece over if you wish."

I experimented for a while. Suddenly, the answer appeared before me. "Soames! They make a square—Beatrix's motif! How beautiful!"

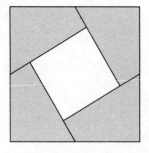

Watsup's arrangement

Soames peered at my little jigsaw. "Indeed it is. Do you still contend that an elegant explanation of how several clues fit together constitutes definitive evidence that you have found the guilty party?"

"How else could the evidence fit so tightly, Soames?"

"How else indeed?" I realised that his question was rhetorical. "There is a hole in your argument, Watsup," he went on, when I declined to answer. "Let us eliminate it." He reached down and rearranged the pieces to form a complete square.

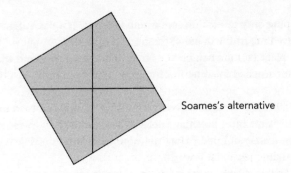

Soames's alternative

"Oh," said I, shamefaced. "*That* is Beatrix's motif, then."

"So I conjecture. But be not downcast: your arrangement is Miss Makepiece's."

Light dawned. "You think that copies of this one tile can form all thirteen motifs?"

"I am certain of it. See: here is how three tiles form Mrs Wotton's motif, an equilateral triangle with a triangular hole."

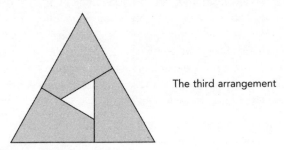

The third arrangement

"Good heavens, Soames!"

"It is a remarkably versatile—er—tile," he replied. "Thanks to its cunning geometry."

"So all we have to do—" I began.

"—is to find arrangements that fit the other ten motifs!" Roulade finished for me.

Soames began to ream out his pipe. "I am sure I can safely leave that to you gentlemen."

That evening, I took a cab to Beatrix's father's house,

stopping only to pick up something from the jeweller's. She received me in the drawing-room.

I placed a long box on the table. "Open it, my dear."

She reached for it hesitantly, hope dawning on her lovely face.

"Oh! John, you have recovered my pendant!" She took my hand. "How can I ever thank you?" Suddenly she fell silent. "But—*this* is not mine." She reached into the box and picked out a sparkling jewel. "It is an engagement ring."

"So it is. And it *can* be yours," said I, going down on one knee.

Can you find the other ten arrangements? See page 292 for the answers.

Can you find the other ten arrangements? See page 292 for the answers.

* *

The Thrackle Conjecture

A graph is a collection of dots (nodes) joined by lines (edges). When a graph is drawn in the plane, the edges often cross. In 1972 John Conway defined a *thrackle* to be a graph drawn in the plane for which any two edges meet at a node and otherwise do not cross, or do not meet at a node but cross exactly once. The name is said to have been inspired by a Scottish fisherman complaining that his line was 'thrackled'—tangled.

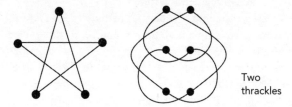

Two thrackles

The figure shows two thrackles. The left-hand one has 5 nodes and 5 edges, while the right hand one has 6 nodes and 6 edges. Conway conjectured that, for any thrackle, the number of edges is less than or equal to the number of nodes. He offered a prize of a bottle of beer for a proof or disproof, but as the years passed and no one solved the riddle, the prize rose to $1,000.

Both thrackles shown are closed loops (nodes on a circle), drawn with overlaps. It is known that any closed loop with $n \geq 5$ nodes can be drawn to form a thrackle. If so, the number E of edges can be equal to the number n of nodes whenever $n \geq 5$. Paul Erdős proved that the conjecture is true for any graph with straight edges. The best bound on the size of E was proved by Radoslav Fulek and János Pach in 2011:

$$E \leq \frac{167}{117}n$$

For further information see page 292.

For further information see page 292.

• •

Bargain with the Devil

A mathematician, who has spent ten fruitless years trying to prove the Riemann Hypothesis, decides to sell his soul to the Devil in exchange for a proof. The Devil promises to show him the proof in a week, but nothing happens.

A year later, the Devil turns up again, looking gloomy. "Sorry, couldn't prove it either," he says, handing back the mathematician's soul. He pauses, and his face lights up. "But I think I found a really interesting lemma ... "

At risk of spoiling the joke, I'd better explain that in mathematics a lemma is a minor proposition whose main interest is to be a potential stepping stone towards something interesting enough to deserve being called a theorem. There is no logical difference between a theorem and a lemma, but psychologically the word 'lemma' indicates that what has been proved goes only part way towards what is really required—

I'll get me coat.

• •

A Tiling That Is Not Periodic

Many different shapes tile the plane without leaving gaps or overlapping. The only regular polygons that can do this are the equilateral triangle, the square, and the hexagon.

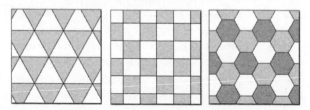

The three regular polygons that can tile the plane

A huge range of less regular shapes also tile the plane, such as the seven-sided polygon in the next picture. It is obtained from a regular heptagon by flipping three of its sides over across the line joining their ends.

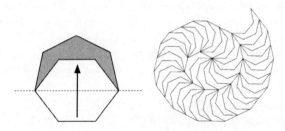

Left: How to make the seven-sided polygon tile from a regular heptagon. *Right*: Spiral tiling.

The regular polygon tilings are *periodic*, that is, they repeat indefinitely in two different directions, like wallpaper patterns. The spiral tiling is not periodic. However, the seven-sided polygon that occurs there *can* also tile the plane periodically.

How can this be done? See page 293 for the answer.

Are there tiles that can tile the plane, but cannot do so periodically? This question has deep connections with mathematical logic. In 1931 Kurt Gödel proved that there exist undecidable problems in arithmetic: statements for which no algorithm can decide whether they are true or false. (An algorithm is a systematic process that is guaranteed to stop with the correct answer.) His theorem implies another, more dramatic one: there are statements in arithmetic that can neither be proved nor disproved.

His example of such a statement was rather contrived, and logicians wondered whether more natural problems might also be undecidable. In 1961 Hao Wang was thinking about the domino problem: given a finite number of shapes for tiles, is there an algorithm that can decide whether or not they can tile the plane? He showed that if there exists a set of tiles that can tile the plane, but cannot tile it periodically, there is no such algorithm. His idea was to encode rules of logic into the shapes of the tiles and use results like Gödel's. It worked, too: in 1966 Robert Berger found a set of 20,426 such tiles, proving that the domino problem is indeed undecidable.

Twenty thousand different shapes is rather a lot. Berger reduced the number to 104; then Hans Läuchli got it down to 40. Raphael Robinson reduced the number of shapes to six. Roger Penrose's discovery in 1973 of what we now call Penrose tiles (see *Cabinet* page 116) reduced the number still further, to a mere two. This left an intriguing mathematical mystery: is there a *single* shape of tile that can tile the plane, but cannot tile it periodically? (The tile's mirror image can be used as well.) The answer was found in 2010 by Joshua Socolar and Joan Taylor [An aperiodic hexagonal tile, *Journal of Combinatorial Theory Series A* 118 (2011) 2207–2231], and it is 'yes'.

Their tile is shown in the figure. It is a 'decorated hexagon', with extra 'matching rules', and it differs from its mirror image. The decorations have to fit together as shown.

Four copies of the
Socolar–Taylor tile
illustrating matching rules

The next figure shows the central region of a tiling of the
plane. You can see that it doesn't look periodic. The paper
explains why this tiling can be continued to cover the entire
plane, and why the result *cannot* be periodic. See their paper for
details.

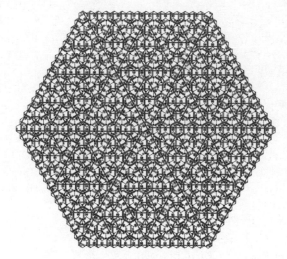

Central region of a tiling of the plane by Socolar–Taylor tiles

The Two Colour Theorem 🔍

From the Memoirs of Dr Watsup

"Well, Soames, here is a jolly little conundrum to brighten your mood." I tossed the *Daily Reporter* across to my friend and companion, the nearly famous detective, who was currently suffering from a fit of depression because his competitor across the street was decidedly more famous, and likely to remain that way.

He tossed it back with a sneer. "Watsup, I have insufficient energy to *read*."

"Then I shall read it to you," I replied. "It seems that the celebrated mathematician Arthur Cayley has published an article in the *Proceedings of the Royal Geographical Society*, asking whether—"

"Whether a map may be coloured with at most four colours so that adjacent regions receive different colours," Soames interrupted. "It is a long-standing problem, Watsup, and I fear it will not be answered in our lifetimes." I said nothing, hoping to draw him out, since this was the longest sentence he had uttered for the best part of a week. My ploy worked, for after an awkward silence he continued: "A young man named Francis Guthrie posed the problem two years before I was born. Unable to solve it, he employed the good offices of his brother Frederick, a student of Professor Augustus De Morgan."

"Ah, yes, Gussie," I interjected, having had some acquaintance with the family of this admirable eccentric, author of *A Budget of Paradoxes* and scourge of mathematical crackpots everywhere.

"De Morgan," Soames went on, "made no progress, so he asked the great Irish mathematician Sir William Rowan Hamilton, who gave him short shrift. And there the problem languished until Cayley once again took it up. Though why he chose to publish in that particular journal, I have no idea."

"Possibly," I essayed, "because geographers are interested in maps?" But Soames was having none of it.

"Not in this manner," he huffed. "A geographer will colour

regions of a map according to political considerations. Adjacency would not affect the choice. Why, Kenya, Uganda, and Tanganyika are all adjacent, but on any map of the British Empire all three will be coloured pink."

I admitted that this was so. Our dear Queen would not be amused were it otherwise. "But Soames," I persisted, "it remains an intriguing question. All the more so because no one seems able to answer it."

Soames grunted.

"Let us make the attempt," said I, and quickly drew a map.

"Curious," said Soames. "Why have you made every region circular?"

"Because any region without holes is topologically equivalent to a circle."

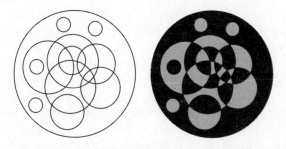

Watsup's map, and how to colour it

Soames pursed his lips. "Even so, it is a poor choice, Watsup."

"Why? It seems to me—"

"Watsup, many things *seem* to you, but few actually are. Although any single region is topologically a circle, two or more regions may overlap in a manner that is impossible for two or more circles. As evidence, your map requires only two colours." He shaded in about half of the regions.

"Well, yes, but I am certain that a more elaborate map of the same kind—"

Soames shook his head. "No, no, Watsup. Any map composed entirely of circular regions, even if they be of different sizes, and overlap in a complicated manner, can be coloured with two colours. Assuming, as is always the case in such questions, that 'adjacent' requires regions to share a common length of border, not just isolated points."

My jaw dropped. "A *two* colour theorem! Astonishing!"

Soames had the grace to shrug. "But how can such a theorem be proved?"

Soames leaned back in his chair. "You know my methods."

See page 293 for the answer.

The Four Colour Theorem in Space

Soames was referring to the celebrated Four Colour Theorem, which tells us that given any map in the plane, its regions can be coloured using at most four distinct colours, so that regions that share a common border have different colours. (Here 'share a border' requires the common border to be of nonzero length; that is, meeting at a single point doesn't count.) This result was conjectured in 1852 by Francis Guthrie, and proved in 1976 by Kenneth Appel and Wolfgang Haken using massive computer assistance.[*] Their proof has since been simplified, but a computer is still essential to carry out a large number of routine but complicated calculations.

Might there be an analogous theorem for 'maps' in space rather than on a plane? Now the regions would be solid blobs. A little thought shows that such a map can require any number of colours. Suppose, for instance, that you want a map needing six colours. Start with six distinct spheres. Let sphere 1 put out five thin tentacles that touch spheres 2, 3, 4, 5, and 6. Then let sphere 2 put out four thin tentacles that touch spheres 3, 4, 5, and 6. Now move on to sphere 3, and so on. Then each tentacled region touches all five others, so they must all have different colours. If

[*] See *Cabinet* (pages 10–16) for a discussion of the problem and its eventual solution.

you did this with 100 spheres, you'd need 100 colours; with a million spheres, you'd need a million colours. In short, there is no limit to the number of colours required.

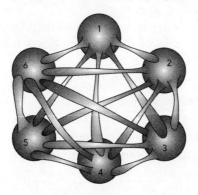

A 'map' in space
needing six colours

In 2013 Bhaskar Bagchi and Basudeb Datta [Higher-dimensional analogues of the map coloring problem, *American Mathematical Monthly* 120 (October 2013) 733–736] realised that this is not the end of the story. Think of 'maps' formed by a finite number of circular discs in the plane that either do not overlap or touch at a common point. Suppose you want to colour the discs so that *touching* discs get different colours. How many do you need? It turns out that again the answer is 'at most four'.

In fact, this problem is essentially equivalent to the Four Colour Theorem. This theorem can be reformulated as the problem of colouring the nodes of a network (or graph, see page 201) in the plane, with no edges crossing each other, so that if two nodes are joined by an edge then the nodes receive different colours. Just create a node for each region of the map and an edge between two nodes if the corresponding regions share a common boundary. It can be proved that any network in the plane can be produced from a suitable set of circles by joining the centres of those circles that touch. For instance, here is a set of circles needing four colours, the associated network, and a map with a topologically equivalent distortion of that network that *also* needs four colours.

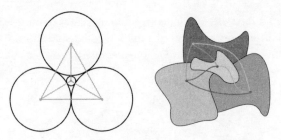

Left: Four circles and the associated network (grey dots and lines). *Right*: A map with a topologically equivalent network, needing four colours.

We can extend the disc formulation naturally to three dimensions, by using spheres instead of discs. Again either the spheres do not overlap or they touch at a common point. Suppose you want to colour the spheres so that touching spheres get different colours. How many do you need? Bagchi and Datta explained why this number must be at least 5 and no greater than 13. The exact number is currently a mathematical mystery. But you may be able to prove that at least five colours are needed. Their result implies that some three-dimensional maps are not equivalent to those obtained from spheres.

See page 295 for the answer.

. . . .

Comical Calculus

For this one you need to know some calculus. If $\int$ is the integration symbol, then the exponential function e^x is its own integral:

$$e^x = \int e^x$$

Therefore

$$(1 - \int)\, e^x = 0$$

so

$$e^x = (1 - \textstyle\int)^{-1} 0$$
$$= (1 + \textstyle\int + \int^2 + \int^3 + \int^4 + \ldots)\, 0$$
$$= 0 + 1 + x + \frac{x^2}{2} + \frac{x^3}{6} + \frac{x^4}{24} + \cdots$$
$$= 1 + x + \frac{x^2}{2!} + \frac{x^3}{3!} + \frac{x^4}{4!} + \cdots$$

The calculation seems to be nonsense; even the first line really ought to be $e^x = \int e^x\, dx$. And a later step takes the formula

$$1 + y + y^2 + y^3 + y^4 + \cdots = (1 - y)^{-1}$$

for summing an infinite geometric series and replaces y by $\int$. This formula is valid when y is a number, less than 1. But $\int$ isn't even a number, just a symbol. How ludicrous!

Despite which, the final result is the *correct* power series for e^x.

This isn't a coincidence. With the right definitions (for example, $\int$ is an operator, transforming a function into its integral, and the 'geometric series' formula works for operators under suitable technical conditions) everything can be made perfectly logical. But it does look strange.

• •

The Erdős Discrepancy Problem

Paul Erdős

Paul Erdős was an eccentric but brilliant Hungarian mathematician. He never owned a house or had a regular academic job, preferring to travel the world with a suitcase and sleep in the homes of understanding colleagues. He published 1,525 mathematical research papers, collaborating with 511 other mathematicians—a figure no one else has come near. He preferred ingenuity to deep systematic theories, and his delight was to solve problems that looked simple, but turned out not to be. His main achievements were in combinatorics, but he could turn his hand to many areas of mathematics. He found a much simpler proof of Bertrand's postulate (there is always a prime between n and $2n$) than the original analytic one of Pafnuty Chebyshev. A high point of his career was a proof of the prime number theorem (the number of primes less than x is approximately $x/\log x$) that avoided complex analysis, previously the only known route to a proof.

He had a lifelong habit of offering monetary prizes for solutions to problems that he came up with but was unable to solve himself. He would offer $25 for a solution of something he suspected to be relatively easy, and thousands of dollars for something he believed was really hard. A typical example of his kind of mathematics is the Erdős Discrepancy Problem, priced at $500. It was posed in 1932 and solved early in 2014. It is a remarkable example of how today's mathematicians approach long-standing mysteries.

The problem starts with an infinite sequence of numbers, each either $+1$ or -1. It might be a regular sequence, such as

$$+1 \ -1 \ +1 \ -1 \ +1 \ -1 \ +1 \ -1 \ +1 \ -1 \ \ldots$$

or an irregular ('random') one such as

$$+1 \ -1 \ -1 \ -1 \ +1 \ -1 \ +1 \ +1 \ -1 \ +1 \ \ldots$$

which I got by tossing a coin. It need not contain the same proportion of $+$ and $-$ signs. *Any* sequence will do.

One way to see that the first of these is regular is to look at every second term:

$$-1 \ -1 \ -1 \ -1 \ -1 \ \ldots$$

The sums of the first n terms go

$$-1 \ -2 \ -3 \ -4 \ -5 \ \ldots$$

and decrease without limit. If we do the same for the second sequence we get

$$+1 \ -1 \ -1 \ +1 \ -1 \ \ldots$$

with sums

$$+1 \ 0 \ -1 \ 0 \ +1 \ \ldots$$

that go up and down.

Suppose we take a specific but arbitrary sequence of ± 1's, and set ourselves a positive target number C, which can be as large as we wish—a billion, say. Erdős asked whether there is always some number d such that the sums of terms d steps apart, at positions d, $2d$, $3d$, and so on, either become larger than C or smaller than -C at some stage.

Having reached the target, they may subsequently give sums between C and -C: it is enough to hit the target once. However, there has to be a suitable step size d for *any* target C. Of course, d depends on C. That is, if the sequence is $x_1, x_2, x_3, \ldots$, can we find d and k so that

$$|x_d + x_{2d} + x_{3d} + \cdots + x_{kd}| > C?$$

The absolute value of the sum on the left is the *discrepancy* of the sub-sequence determined by the step-size d, and it measures the excess of $+$ signs over $-$ signs (or the other way round).

Early in February 2014 Alexei Lisitsa and Boris Konev announced that the answer to Erdős's question is 'yes' if $C = 2$. Indeed, if we select a d-step sub-sequence from the first 1,161 terms of any ± 1 sequence, and choose the appropriate length k, the absolute value of the sum exceeds $C = 2$. Their proof requires heavy use of a computer, and the details require a 13-gigabyte data file. This is more than the entire contents of Wikipedia, at 10 gigabytes. It is certainly one of the longest proofs ever, and already too long for a human being to check.

Lisitsa is now looking for a proof for $C = 3$, but the computer has not yet completed its calculations. It is sobering to think that a complete solution requires understanding what happens for *any* choice of C. The hope is that computer solutions for small C might reveal a new idea, which a human mathematician could turn into a general proof. On the other hand, the answer to Erdős's question might be 'no'. If so, there's a really interesting sequence of ± 1's out there, waiting to be defined.

• •

The Greek Integrator

From the Memoirs of Dr Watsup

Although my friend's investigatory powers are mainly directed towards the pursuit of crime, from time to time his skills are applied in the service of scholarship. One such instance was a singular quest that we carried out in the Autumn of 1881 at the behest of a wealthy but reclusive collector of ancient manuscripts. With the aid of a torn page from an old notebook, a lantern, a bunch of skeleton keys, and a large crowbar, Soames and I located an enormous flagstone and levered it up to uncover a spiral stairway that led down to a concealed chamber deep beneath the library of a famous European University.

Soames consulted the torn scrap of paper, much damaged by fire and water. "The Lost Incunabula of the Cartonari," he explained.

"Them again!" He had mentioned this name in passing during the Adventure of the Cardboard Boxes (see page 23), but declined to say more. Now I pressed him for futher details.

"The name means 'cardboard manufacturers'. It is an Italian secret society organised along the lines of the Freemasons and dedicated to the cause of Nationalism, being implicated in the failed revolution of 1820."

"I recall the revolution with the utmost clarity, Soames. But not the organisation."

"Few indeed are aware of its hidden hand." He consulted the

scrap of paper. "The page is almost obliterated, but it takes no great expertise in higher mathematics to recognise that it is some form of Fibonacci code, rewritten in Da Vinci mirror script, and transformed into a sequence of rational points on an elliptic curve."

"Even a child could see that," said I, lying through my teeth.

"Quite. Now, if I read these runes aright, we should find what we seek somewhere on these shelves."

After a moment, I asked: "Soames, what *do* we seek? You have played your cards unusually close to your chest."

"It is knowledge that holds great dangers, Watsup. I saw no need to expose you to violence prematurely. But now that we have penetrated the inner sanctum—Ah! Here it is!" He extracted what I immediately recognised to be a parchment codex, blowing off centuries of accumulated dust.

"What the devil is that, Soames?"

"Do you have your service revolver?"

"Never without it."

"Then it is safe to tell you that in my hands I hold . . . the Archimedes Palimpsest!"

"Ah."

Now, I was aware that a palimpsest is a document that has been written on and then scrubbed clean to allow further inscription, and that scholars can with difficulty reconstruct that which has been obliterated, thereby retrieving a hitherto unknown gospel from the laundry list of an obscure order of fourteenth-century monks. Archimedes was also familiar to me, as an ancient Greek geometer of prodigious ability. It was therefore apparent that Soames had unearthed some previously unknown mathematical text. But he insisted that we should make our exit immediately, before the Inquisitorial Vengeance Squad descended upon us.

Back in the comparative safety of Baker Street, we inspected the document.

"It is a tenth-century Byzantine copy of a hitherto unknown work of Archimedes," said Soames. "Its title loosely translates as

The Method: it concerns that geometer's celebrated work on the volume and surface area of a sphere. It shows us how he came to discover these results, offering an unparalleled insight into his thought processes."

I was struck dumb, and no doubt resembled a goldfish out of water.

"Archimedes discovered that if a sphere is inscribed in a tightly fitting cylinder, then the volume of the sphere is exactly two thirds the volume of the cylinder, and its area is the same as that of the curved surface of the cylinder. In modern language, if the radius is r then the volume is $\frac{4}{3}\pi r^3$ and the area $4\pi r^2$.

"Now, Archimedes was such a great mathematician that he was able to find a logically rigorous geometric proof of these facts, which he included in his book *On the Sphere and Cylinder*. There he used a complicated method of proof now known as exhaustion. One of its tricky features is that one must know the exact answer to the problem before proving it to be correct. So it has long been a puzzle to scholars: how did Archimedes *know* what the right answer must be?"

"I see," said I. "This long-lost document explains how he did it."

"Exactly. Remarkably, it comes close to being an anticipation, for this particular example, of the integral calculus of Isaac Newton and Gottfried Leibniz, developed more than two thousand years later. But, as Archimedes well knew, the ideas used in *The Method* lack rigour. Hence his use of exhaustion, a very different approach."

"So how *did* he do it?" I asked.

Soames studied the palimpsest through his magnifying glass. "The Greek is not entirely classical, and often unclear, but that poses no serious difficulty to an expert linguist such as myself. Have I shown you my pamphlet on the decipherment of obscure ancient texts from the Mediterranean region? Remind me to do so.

"It seems that Archimedes began with a sphere, a cone, and a cylinder of suitable dimensions. Then he imagined taking a very thin slice of each, and hanging them on a balance: a slice of the sphere and a slice of the cone on one side, a slice of the cylinder on the other. If the distances are chosen correctly, the masses will balance exactly. Since mass is proportional to volume, the volumes are related by the law of the lever."

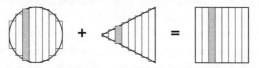

Slicing the solids prior to hanging them from a balance: see page 298 for details

"Uh—pray remind me of this law," said I. "It was unaccountably not part of the syllabus at medical school."

"It should have been," said Soames. "It would be of great use when dealing with dislocated joints. No matter. The law, which was discovered and proved by Archimedes, states that the turning effect, or moment, of a given mass at a given distance is the mass multiplied by the distance. For a set of masses to balance, the total clockwise moment must equal the total anticlockwise moment. Or, with appropriate assignment of plus and minus signs, the total moment must be zero."

"Er—"

"A mass at a given distance will balance half that mass at twice the distance, provided it is *on the other arm of the balance*."

"I see."

"I suspect not, but let me proceed. By splitting these solids into infinitely many infinitely thin slices, and hanging them appropriately on his balance, Archimedes was able to concentrate the entire mass of the sphere, and that of the cone, at a single point. The slices of cylinder, which are identical circles, are placed at different distances; together they reconstruct the original cylinder. Knowing that the volume of the cone, hence its mass, is one third that of the cylinder, Archimedes could then solve the resulting 'equation' for the volume of the sphere."

"Amazing," said I. "It seems convincing enough to me."

"But not to a mathematician of the intellectual calibre of

Archimedes," said Soames. "If the slices have finite thickness, the procedure involves small but unavoidable errors. But if the slices have zero thickness, they have zero mass. There is no unique balance point when the masses concerned are all zero."

I began to see the objection to the procedure. "Presumably the errors become ever smaller as the slices become thinner?" I hazarded.

"They do, Watsup, they do. And the modern approach to integral calculus converts that statement into a proof that this kind of process gives sensible answers. However, these ideas were not available to Archimedes. So he used a non-rigorous method to find the correct answer, and that allowed him to use exhaustion to prove the answer was correct."

"Astonishing," I said. "We must publish the palimpsest."

Soames shook his head. "And risk the wrath of the Cartonari? I value both our lives too highly to attract their attention."

"So what should we do?"

"We must place the manuscript somewhere safe. Not back in the library, for they must by now have noticed its disappearance and set subtle traps. I will conceal it in some *other* scholarly library. No, do not ask which! Perhaps at some later date, when times are less troubled and the influence of secret societies has waned, it will be rediscovered. Until then, we must be content with knowing the great geometer's method, even though we cannot reveal it to the world."

He paused. "I have already told you the formulas for the area and volume of a sphere. So here is a simple little conundrum that might amuse you. What should the radius of the sphere be, in feet, so that its area in square feet is exactly equal to its volume in cubic feet?"

"I have no idea," said I.

"Then *work it out*, man!" he cried.

See page 296 for the true history of the Archimedes palimpsest and the answer to Soames's puzzle.

Sums of Four Cubes

Sums of four *squares*, like many mathematical mysteries, have a long history. The Greek mathematician Diophantus, whose *Arithmetica* of about AD 250 was the first textbook to use a form of algebraic symbolism, asked whether every positive whole number is a sum of four perfect squares (0 allowed). It's easy to verify this statement experimentally for small numbers; for example,

$$5 = 2^2 + 1^2 + 0^2 + 0^2$$
$$6 = 2^2 + 1^2 + 1^2 + 0^2$$
$$7 = 2^2 + 1^2 + 1^2 + 1^2$$

Just when you think that 8 will need another 1^2, hence five squares, 4 comes to the rescue:

$$8 = 2^2 + 2^2 + 0^2 + 0^2$$

Experiments with larger numbers strongly suggest that the answer is 'yes', but the problem remained unsolved for over 1,500 years. It became known as Bachet's problem, after Claude Bachet de Méziriac published a French translation of *Arithmetica* in 1621. Joseph-Louis Lagrange found a proof in 1770. Simpler proofs have been found more recently, based on abstract algebra.

What about sums of four *cubes*?

Also in 1770, Edward Waring stated without proof that every positive whole number is the sum of at most 9 cubes and 19 fourth powers, and asked whether similar statements are true for higher powers. That is, given a number k, is there some finite limit to the number of kth powers needed to express any positive whole number by adding them up? In 1909 David Hilbert proved that the answer is 'yes'. (Odd powers of negative numbers are negative, and that changes the game considerably, so for the moment we restrict attention to powers of positive numbers.)

The number 23 definitely requires 9 cubes. The only possibilities are 8, 1, and 0, and the best we can do is $8 + 8 +$ seven 1's:

$$23 = 2^3 + 2^3 + 1^3 + 1^3 + 1^3 + 1^3 + 1^3 + 1^3 + 1^3$$

So we can't succeed in general with fewer than 9 cubes. However, that number can be reduced if we ignore a finite number of exceptions. For example only 23 and 239 actually need 9 cubes; all others can be achieved using 8. Yuri Linnik reduced this to 7 by allowing a few more exceptions, and it is widely believed that the correct number, allowing finitely many exceptions, is 4. The largest known number that needs more than 4 cubes is 7,373,170,279,850, and it is conjectured that no larger numbers with that property exist. So it is very likely—but remains an open question—that every sufficiently large positive whole number is a sum of four positive cubes.

But, as I said earlier, the cube of a negative number is negative. This allows new possibilities that do not occur for even powers. For example,

$$23 = 27-1-1-1-1 = 3^3 + (-1)^3 + (-1)^3 + (-1)^3 + (-1)^3$$

using only 5 cubes, whereas with positive or zero cubes it requires 9, as we've just seen. But we can do better: 23 can be expressed using just four cubes:

$$23 = 512 + 512 - 1 - 1000 = 8^3 + 8^3 + (-1)^3 + (-10)^3$$

Allowing negative numbers means that the cubes involved might be much larger (ignoring the minus sign) than the number concerned. As an example, we can write 30 as a sum of three cubes, but we have to work pretty hard:

$$30 = 2{,}220{,}422{,}932^3 + (-283{,}059{,}965)^3 + (-2{,}218{,}888{,}517)^3$$

So we can't work systematically through a limited number of possibilities, as we can when only positive numbers are considered.

Experiments have led several mathematicians to conjecture that *every* integer is the sum of four (positive or negative) integer cubes. As yet, this statement remains mysterious, but the evidence is substantial. Computer calculations verify that every positive integer up to 10 million is a sum of four cubes. V. Demjanenko has proved that any number not of the form $9k \pm 4$ is always a sum of four cubes.

Why the Leopard Got Its Spots

Leopardess, Kanana Camp, Botswana

Leopards have spots, tigers have stripes, and lions are plain. Why? It all seems rather arbitrary, as if the Big Cat Sales Catalogue lists coat patterns and evolution picks whichever looks prettiest. But evidence is accumulating that it's not like that. William Allen and colleagues have investigated how the mathematical rules that determine the patterns relate to the cats' habits and habitats, and how this affects the way the patterns evolve.

The most obvious reason for evolving patterned coats is camouflage. If a cat lives in the forest, spots or stripes will make it hard to see among the light and shade. Cats that operate out in the open, on the other hand, will be *easier* to see if they have strong patterns. However, theories of this kind are little better than just-so stories, unless they can be supported by evidence. Experimental verification is difficult: imagine painting out a tiger's stripes for several generations, or fitting it and its descendants with plain overcoats, to see what happens.

Alternative theories abound: markings might exist to attract mates, or merely be a natural consequence of the animal's size.

The mathematical model of cat patterning makes it possible to test the camouflage theory. Some patterns, such as the leopard's spots, are very complex, and the type of complexity is closely related to the pattern's value as camouflage. So the researchers classified patterns using a mathematical scheme invented by Alan Turing, in which the pattern is laid down by chemicals that react together and diffuse across the surface of the developing embryo.

These processes can be characterised by specific numbers that determine the rate of diffusion and the type of reaction. These numbers act like coordinates on 'camouflage space', the set of all possible patterns, just as latitude and longitude provide coordinates on the surface of the Earth.

The research relates these numbers to observational data on 35 different cat species: what kind of habitat the cats prefer, what they eat, whether they hunt by day or by night. Statistical methods identify significant relationships between these variables and the animals' coat patterns. The results show that patterns are closely associated with closed environments, such as forests. Animals in open environments, such as savannahs, are more likely to be plain, like lions. If not, they usually have simple patterns. But cats that spend a lot of time in trees, such as leopards, are more likely to have patterned coats. Moreover, these patterns tend to be complex, not just simple spots or stripes. The method also explains why black leopards ('panthers') are common, but there are no black cheetahs.

The data argue against several alternatives to camouflage. The size of the cat and the size of its prey have little effect on patterns. Sociable cats are no more or less likely to be patterned than solitary ones, so the markings are probably not important for social signalling. This work does not prove that cats' markings evolved for camouflage alone, but it suggests that camouflage played a key evolutionary role.

Lions are plain because they prowl on the plains. Leopards are spotty because spots are harder to spot.

See page 299 for further information.

● ●

Polygons Forever

Keep going forever... How big does it get?

Here's a test of your geometric and analytic intuition. Start with a circle of unit radius. Draw the tightest fitting equilateral triangle that you can round it; then draw the tightest fitting circle round that. Repeat, but at successive stages use a square, a regular pentagon, a regular hexagon, and so on.

If this process goes on forever, does the picture become arbitrarily large, or does it remain within a bounded region of the plane?

See page 299 for the answer.

● ●

Top Secret

In the 1930s a Russian mathematics professor was running a seminar in fluid dynamics. Two of the regular attendees, always dressed in uniform, were obviously military engineers. They never discussed the project they were working on, which was presumably top secret. But one day they asked the professor for help with a mathematical problem. The solution of a certain equation oscillated, and they wanted to know how to change the coefficients to make it monotonic.

The professor looked at the equation and said: "Make the wings longer!"

. . . .

The Adventure of the Rowing Men

From the Memoirs of Dr Watsup

I have often been astonished by Soames's ability to perceive patterns in the most unpromising of circumstances. No better example could be found than that which occurred in the early spring of 1877.

As I walked across Equilateral Park towards his lodgings, a freshly minted sun cast dappled light and shade through a scattering of fluffy clouds, and the hedgerows rang with birdsong. On so glorious a day it seemed veritably indecent to remain indoors, but my efforts to drag my friend away from cataloguing his comprehensive collection of used matchsticks met with indifference.

"Many a case has hinged upon the time it took a match to burn, Watsup," he grumbled, transferring a measurement from his dividers into a notebook.

Disappointed, I opened the newspaper at the sports pages, and my eye was caught by a timely reminder of an event that even Soames would not wish to miss. It had completely escaped my mind amid the buzzing of the bees and the blossoming trees. Within the hour the two of us were seated on the riverbank with a luncheon basket and several bottles of a palatable Burgundy, awaiting the start of the annual race.

"Whom do you favour, Soames?"

He stopped measuring the length of the burn mark on an early Scottish lucifer, for he had insisted on bringing a number of matchsticks with him to help pass the time. "The blue team."

"Dark, or light?"

"Yes," he said enigmatically.

"I mean: Oxford or Cambridge?"

"Yes." He shook his head. "One of those. The variables are too complex to make a prediction, Watsup."

"Soames, my query was about support, not prediction."

He gave me a scathing look. "Watsup, why should I support men with whom I am not acquainted?"

When Soames is in a mood, there is always a reason. I noticed that he was laying out matchsticks in patterns resembling the bones on a kipper, and asked him why.

"I have been observing the distribution of the oars, and I am wondering why such an inefficient arrangement has become traditional."

I looked at the two boats as they lined up on the Thames for the annual University Boat Race. "Tradition is often inefficient, Soames," I chided. "For it consists of doing things the same way they have always been done, instead of asking how best to do them. But I see no inefficiency here. There are eight rowers, and the oars point alternately to the left and the right. It is known as tandem rigging. That seems symmetric and sensible to me."

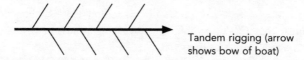

Tandem rigging (arrow shows bow of boat)

Soames gave a dissatisfied grunt. "Symmetric? Pah! Not at all. The oars on one side of the boat are all in front of the corresponding oars on the other side. Sensible? The asymmetry creates a twisting force when the rowers pull on the oars, causing the boat to veer to one side."

"That, Soames, is one reason why there is a coxswain. Who has a rudder."

"Which creates resistance to the forward motion of the boat."

"Oh. But how else could the oars be arranged? It is not possible to sit two rowers side by side."

"There are 68 alternatives, Watsup; 34 if we count left–right reflections as being the same. Our German and Italian friends use different arrangements, to be specific." He laid out two skeletal arrangements of matchsticks.

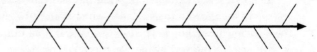

Left: German rigging. *Right*: Italian rigging.

I stared at them. "Surely such strange arrangements suffer from even worse problems!"

"Perhaps. Let us see." He pursed his lips, deep in thought. "There are innumerable practical issues, Watsup, which would require a more complex analysis. Not to mention more matchsticks. So I shall content myself with the simplest model I can devise, in the hope of gaining useful insights. I warn you now, the results will not be definitive."

"Fair enough," said I.

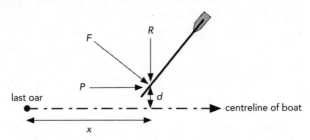

Resolving the forces. Note that P points *forwards* and R points away from the centre of the boat because the outermost end of the oar is held fixed (to a good approximation) by the resistance of the water. Don't forget that the rower faces the stern and pulls the oar towards him.

"Let us then consider a single oar, and calculate the forces acting on the rowlock where it pivots, during that part of the stroke when the oar is in the water. For simplicity I shall assume that all rowers have the same strength, and row in perfect synchrony, so they exert identical forces F at any given instant. I then resolve this force into a component P parallel to the axis of the boat, and R at right angles."

"All of these forces vary with time," said I.

He nodded. "What matters is what mechanicians call the *moment* of each force—the extent to which it turns the boat about some chosen point. This, you will recall from our encounter with the Archimedes Palimpsest, is found by multiplying the force by its perpendicular distance from that point."

It was my turn to nod. I was sure I had remembered something of the kind.

"I mark the position of the sternmost oar by a dot. This will be our chosen point. Now, the force P has moment Pd about the point at which the rowlock meets the central axis of the boat, if the oar is on the left-hand side. But if it is on the right-hand side, the moment is $-Pd$ since the force acts in the opposite direction. Notice that these moments are the same for all four oars on the same side of the boat. In consequence, the total moment of all eight oars is $4Pd-4Pd$, which is 0."

"The twisting forces cancel out!"

"For the parallel forces P, yes. However, the moment of the force R is different for each oar, for it depends on the distance x between that oar and the one at the stern. In fact, it is Rx. If successive oars are separated by the same distance c, then x takes the values

$$0 \quad cR \quad 2cR \quad 3cR \quad 4cR \quad 5cR \quad 6cR \quad 7cR$$

as the oars run from the stern to the bow. Therefore the total moment is

$$\pm 0 \pm cR \pm 2cR \pm 3cR \pm 4cR \pm 5cR \pm 6Rc \pm 7cR$$

where the sign is plus for oars on the left of the boat, but minus for those on the right."

"Why?"

"Forces on the left side turn the boat clockwise, Watsup,

whereas those on the right turn it anticlockwise. We can simplify this expression to

$$(\pm 0 \pm 1 \pm 2 \pm 3 \pm 4 \pm 5 \pm 6 \pm 7)cR$$

where the pattern of plus and minus signs matches the sequence of sides at which the oars are placed.

"Now consider tandem rigging. Here the sequence of signs is

$$+ - + - + - + -$$

so the combined turning moment is

$$(0 - 1 + 2 - 3 + 4 - 5 + 6 - 7)cR = -4cR$$

During the first part of the stroke, R points inwards, but once the oar starts to trail backwards the direction of R reverses and it points outwards. So the boat first turns in one direction, then in the other, creating a wiggling motion. The cox has to use the rudder to correct this, and—as I have said—this creates resistance.

"What of the German rigging? Now the combined turning moment is

$$(0 - 1 + 2 - 3 - 4 + 5 - 6 + 7)cR = 0$$

whatever c and R might be. So there is *no* tendency to wiggle."

"What of the Italian?" I cried. "Oh, do let me try! The combined turning moment is

$$(0 - 1 - 2 + 3 + 4 - 5 - 6 + 7)cR = 0$$

as well! How remarkable."

"Quite," replied Soames. "Now, Watsup, here is a question for your agile mind. Are the German and Italian rigs—or their left–right reflections, which differ only trivially—the *only* ways to make the turning forces equal zero?" He must have seen the look on my face, for he added: "The question boils down to separating the numbers from 0 to 7 into two sets of four, each having the same sum. Which must be 14 since all seven numbers add up to 28."

See page 300 for the answer, and for the result of the 1877 Boat Race.

• •

The Fifteen Puzzle

This is an old favourite, but none the worse for that. It's a fascinating case where a little mathematical insight could have saved an awful lot of wasted effort. Plus, I need it to set up the next item.

In 1880 a New York postmaster named Noyes Palmer Chapman came up with what he called the Gem Puzzle, and the dentist Charles Pevey offered money for a solution. This triggered a brief craze, but no one won the cash, so it quickly died down. The American puzzlist Sam Loyd* claimed that he had started a craze for this puzzle in the 1870s, but all he really did was to write about it in 1896, offering a prize of $1,000, which revived interest for a time.

The puzzle (also called the Boss Puzzle, Game of Fifteen, Mystic Square, and Fifteen Puzzle) starts with 15 sliding blocks numbered 1–15 arranged in a square, with an empty square at bottom right. Blocks are in numerical order, *except* for 14 and 15. Your task is to swap the 14 and 15, leaving everything else as it was. You do this by moving any adjacent block into the empty square, and repeating these moves as often as you wish.

As you move more and more blocks, the numbers get jumbled up. But if you're careful, you can unjumble them again. It's easy to assume that any arrangement can be obtained if you're clever enough.

Fifteen puzzle. *Left*: Start. *Middle*: Finish. *Right*: Colouring the blocks for an impossibility proof.

---

* That's not a typo: he didn't spell it with a double L.

Loyd was happy to offer such a generous (at the time) prize, because he was confident he would never have to pay up. There are 16! potentially possible arrangements: all possible permutations of the blocks (15 numbered plus one empty). The question is: which of these arrangements can be reached by a series of legal moves? In 1879 William Johnson and William Story proved that the answer is exactly half of them; and—wouldn't you just know it—the arrangement that gets the cash is in the other half. The Fifteen Puzzle is insoluble. But most people didn't know that.

The impossibility proof involves colouring the squares like a chessboard, as in the right-hand figure. Sliding a block in effect swaps that block with the empty square, and each swap changes the colour associated with the empty square. Since the empty square must end up in its original position, the number of swaps must be even. Every permutation can be obtained from some series of swaps; however, half of them use an even number of swaps and the other half use an odd number.

There are many ways to achieve any given permutation, but they are either all odd or all even. The desired result could be obtained with just one swap, interchanging 14 and 15, but one is odd, so you can't achieve this permutation with an even number of swaps.

This condition turns out to be the only obstacle: legal moves lead to exactly half of the 16! possible rearrangements. Now 16!/2 = 10,461,394,944,000; such a large number that however many times you try, most possibilities remain unexplored. This could encourage you to think that *any* arrangement must surely be possible.

● ●

The Tricky Six Puzzle

In 1974 Richard Wilson generalised the Fifteen Puzzle and proved a remarkable theorem. He replaced the sliding blocks by a network. The blocks are represented by numbers that can be slid along an edge, provided it is connected to the node currently bearing the blank square. The blank square then moves to a new location. The figure shows the starting position of the blocks. Nodes are linked if the corresponding squares are adjacent.

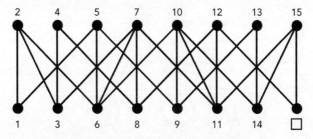

Network representing the Fifteen puzzle

Wilson's idea is to replace this network by any connected network. Suppose it has $n + 1$ nodes. Initially, one node, marked by the box, is empty (from now on, think of it as node 0) and the rest have a number $(1-n)$ sitting on them. The puzzle amounts to moving these numbers around the network, by swapping 0 with the number on an adjacent node. The rules specify that 0 must end up where it started. The other n numbers can be permuted in $n!$ ways. Wilson asked: what fraction of these permutations can be achieved by legal moves? The answer clearly depends on the network, but not as much as you might expect.

There is one obvious class of networks for which the answer is unusually small. If the nodes form a closed ring, the initial arrangement is the *only* one you can reach by legal moves, because 0 has to return to its starting point. All the other numbers stay in the same cyclic order; there's no way for a number to work its way round the side of another one. Rick Wilson's Theorem (so named to avoid confusion with another mathematical Wilson) states that aside from closed rings, either *all* permutations can be achieved, or exactly half of them (the even ones).

With exactly one glorious exception.

The theorem reveals a surprise. A *unique* surprise: a network with seven nodes. Six form a hexagon, and the other one sits in the middle of a diameter. There are 6! = 720 permutations; half of this is 360. But the actual number that can be reached is only 120.

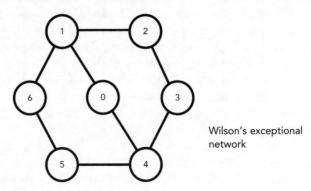

Wilson's exceptional network

The reasoning involves abstract algebra, namely some elegant properties of groups of permutations. Details are in: Alex Fink and Richard Guy, Rick's tricky six puzzle: S5 sits specially in S6, *Mathematics Magazine* 82 (2009) 83-102.

• •

As Difficult as ABC

From time to time mathematicians have apparently crazy ideas that turn out to have huge implications. The ABC Conjecture is one of them.

Remember Fermat's Last Theorem? In 1637 Pierre de Fermat conjectured that if $n \geq 3$ there are no nonzero integer solutions to the Fermat equation

$$a^n + b^n = c^n$$

On the other hand, there are infinitely many solutions when $n = 2$, such as the Pythagorean triple $3^2 + 4^2 = 5^2$. It took 358 years to prove Fermat was right, with the work of Andrew Wiles and Richard Taylor (see *Cabinet* page 50).

Job done, you might think. But in 1983 Richard Mason realised that no one had looked closely at Fermat's Last Theorem for *first* powers:

$$a + b = c$$

You don't need to be a whizz at algebra to find solutions: $1 + 2 = 3$, $2 + 2 = 4$. But Mason wondered whether the question gets more interesting if you impose deeper conditions on a, b, and c. What eventually emerged was a shining new conjecture, the ABC (or Oesterlé–Masser) Conjecture, which will revolutionise number theory if anyone can prove it. It's supported by a vast quantity of numerical evidence, but a proof seems elusive, with the possible exception of work of Shinichi Mochizuki. I'll get back to that once we know what we're talking about.

More than two thousand years ago, Euclid knew how to find all Pythagorean triples, using what we would now write as an algebraic formula. In 1851 Joseph Liouville proved that no such formula exists for the Fermat equation when $n \geq 3$. Mason wondered about the simpler equation

$$a(x) + b(x) = c(x)$$

where $a(x)$, $b(x)$, and $c(x)$ are polynomials. A polynomial is an algebraic combination of powers of x, like $5x^4 - 17x^3 + 33x - 4$.

Again, it's easy to find solutions, but they can't all be 'interesting'. The degree of a polynomial is the highest power of x that occurs. Mason proved that if this equation holds, the degrees of a, b, and c are all less than the number of *distinct* complex solutions x of the equation $a(x)b(x)c(x) = 0$. It turned out that W. Wilson Stothers had proved the same thing in 1981, but Mason developed the idea further.

Number theorists often look for analogies between polynomials and integers. The natural analogue of the Mason–Stothers theorem would be this. Suppose that $a + b = c$, where a, b, and c are integers with no common factor. Then the number of prime factors of each of a, b, and c is less than the number of *distinct* prime factors of *abc*.

Unfortunately, this is spectacularly false. For example,

$9 + 16 = 25$, which has only one (necessarily distinct) prime factor, whereas $9 = 3 \times 3$ has two prime factors and $6 = 2 \times 2 \times 2 \times 2$ has four. Oops. Undaunted, mathematicians tried to modify the statement until it looked like it might be true. In 1985 David Masser and Joseph Oesterlé did just that. Their version states:

> For every $\varepsilon > 0$, there are only finitely many triples of positive integers, with no common factors, satisfying $a + b = c$, such that $c > d^{1+\varepsilon}$, where d denotes the product of the *distinct* prime factors of *abc*.

This is the ABC Conjecture. If it were to be proved, many deep and difficult theorems, proved over past decades, with enormous insight and effort, would be direct consequences, and therefore have simpler proofs. Moreover, all of those proofs would be very similar: do some minor and routine setting up, and then apply the ABC *Theorem*, as it would then become. Andrew Granville and Thomas Tucker [It's as easy as *abc*, *Notices of the American Mathematical Society* 49 (2002) 1224–1231] write that a resolution of the conjecture would have "an extraordinary impact on our understanding of number theory. Proving or disproving it would be amazing."

Back to Mochizuki, a well-respected number theorist with a solid track record of research. In 2012 he announced a proof of the ABC Conjecture in a series of four preprints—papers not yet submitted for official publication. Contrary to his intentions, this attracted the attention of the media, though it was surely unrealistic to imagine that this could possibly have been avoided. Experts are currently checking the 500 or so pages of radically new mathematics involved. This is taking a lot of time and effort because the ideas are technical, complicated, and unorthodox; however, no one is rejecting it because of that. One error has been found, but Mochizuki has stated that it does not affect the proof. He continues to post progress reports, and the experts continue checking.

Rings of Regular Solids

Eight identical cubes fit together, face to face, to make a cube twice the size. Eight cubes also fit together to form a 'ring'—a solid with a hole in it, topologically a torus.

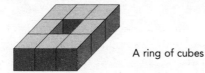

A ring of cubes

With a bit more effort, you can do the same with three other regular solids: the octahedron, dodecahedron, and icosahedron. In all four cases the solids are exactly regular and they fit together exactly: this is obvious for cubes and is a simple consequence of symmetry for the other three solids.

Rings of octahedrons, dodecahedrons, and icosahedrons

However, there are *five* regular solids, and this method doesn't work for the one remaining type, a tetrahedron. So in 1957 Hugo Steinhaus asked whether a number of identical regular tetrahedrons can be glued face-to-face so that they form a closed ring. His question was answered a year later, when S. Świerczkowski proved that such an arrangement is impossible. Tetrahedrons are special.

However, in 2013 Michael Elgersma and Stan Wagon discovered a beautiful eightfold symmetric ring made from 48 tetrahedrons. Was Świerczkowski wrong?

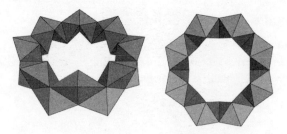

Elgersma and Wagon's ring. *Left*: Perspective view. *Right*: View from above to show eightfold symmetry.

Not at all, as Elgersma and Wagon explained in their article about their discovery. If you make their arrangement using genuinely regular tetrahedrons, they leave a small gap. You can close the gap by elongating the edges shown as thicker lines, from 1 unit to 1.00274, a difference of one part in five hundred, which the human eye can't detect.

The gap, exaggerated

Świerczkowski asked : if you use enough tetrahedrons, and fit them together in a ring to leave a gap, how small can the gap be? Can you make it as small as you please, relative to the size of a single tetrahedron, by using sufficiently many? The answer is still not known if the tetrahedrons are not allowed to intersect, but Elgersma and Wagon prove that the answer is 'yes' if they can interpenetrate. For example, 438 of them leave a gap that is about one part in ten thousand.

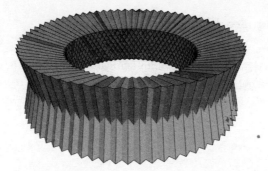

Elgersma and Wagon's 438 interpenetrating tetrahedrons

They conjecture that the answer remains 'yes' even when the tetrahedrons are not allowed to intersect, but the arrangements have to be more complicated. As evidence, they have found a series of rings with ever-narrower gaps. The current record, discovered in 2014, is an almost-closed ring of 540 non-intersecting tetrahedrons with gap 5×10^{-18}.

See page 302 for further information.

Elgersma and Wagon's ring of 540 non-intersecting tetrahedrons.

The Square Peg Problem

This mathematical mystery has been open for more than a century. Is it true that every simple closed curve in the plane (one that does not cross itself) contains four points that are the corners of a square whose side is not zero?

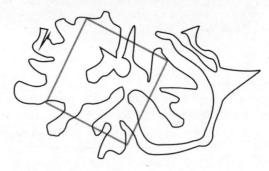

A simple closed curve and a square whose corners lie on it

'Curve' here implies that the line is continuous, no breaks, but it need not be smooth. It can have sharp corners, and indeed it can be infinitely wiggly. We insist that the side of the square is nonzero to avoid the trivial answer: choose the same point for all four corners.

The first printed reference to the Square Peg Problem appeared in 1911 in a report on a conference talk by Otto Toeplitz that apparently claimed a proof. However, no proof was published. In 1913 Arnold Emch proved the statement is true for smooth convex curves, saying that he hadn't heard about it from Toeplitz, but from Aubrey Kempner. The statement has been proved correct for convex curves, analytic curves (defined by convergent power series), sufficiently smooth curves, curves with symmetry, polygons, curves with no cusps and bounded curvature, star-shaped twice-differentiable curves that meet every circle in four points...

You get the picture. Lots of technical hypotheses, no general proof, no counterexample. Perhaps it's true, perhaps not. Who knows?

There are generalisations. The Rectangular Peg Problem asks whether for any real number $r \geq 1$, every smooth simple closed curve in the plane contains the four vertices of a rectangle with sides in the ratio $r : 1$. Only the square peg case $r = 1$ has been proved. There are a few extensions to higher dimensions under very strong conditions.

* *

The Impossible Route

From the Memoirs of Dr Watsup

It is with a heavy heart...

I threw down my pen, overcome with grief. *That Devil's spawn!* Professor Mogiarty's machinations had caused the premature demise of one of the greatest detectives ever to have limped the streets of London disguised as an elderly Russian fishmonger. The finest mind I have ever encountered, snuffed out by a criminal who—until Soames dispatched him at such cost!—had a finger in every foul deed in the kingdom. Except for the idiot who often parks his carriage directly beneath our window, where his horse—

Please bear with your humble scribe as he wipes away a manly tear to recount the tragic events.

Soames had been in a black mood for a week. It was when I saw him fitting the sixth padlock to the window and lining up the third Gatling gun that I began to suspect his mind was troubled in some manner.

"You might say that," said he. "So would yours be if you had narrowly dodged a falling grand piano on your way to the barber's—a Chickering, by the way, I could tell instantly from the cast-iron frame. Before I had gathered my wits, I was forced to leap aside from a runaway brewer's dray drawn by four carthorses, which exploded a split second after I had had the foresight to take cover behind a convenient wall. The prompt collapse of the wall into a deep cavity came close to disturbing what little equilibrium remained to me, but I managed to swing myself to safety using a grappling-iron that I habitually carry in my pocket for such eventualities. It folds up for convenience, and the cord is light but strong. After that, things became a trifle fraught."

If I had not known my friend better, I would have thought he was rattled.

"Has it occurred to you, Soames, that perhaps someone is trying to do you harm?"

He snorted in admiration at my astuteness, or so I presumed. "It is Mogiarty," he stated flatly. "But this time I have the measure of him. Even as we speak my cunning plan is coming to fruition, and every policeman in London is descending on that . . . Wellington of crime . . . and his minions. Soon they will be behind bars, and then . . . the rope!"

There was a knock at the door. An urchin appeared. "Telegram fer 'is nibs!" Soames took the paper and handed the urchin a thruppeny bit.

"The going rate is sixpence," said the urchin.

"Who says so?"

" 'im across the road, guv. That Mr Sher—"

"It'll be tuppence and a clip round the ear if you don't go away," said Soames. The urchin left, muttering under his breath. Soames opened the folded paper. "No doubt news of the operation's succe . . . " His voice trailed off.

"What is it?" I asked anxiously. His face had gone deathly pale.

"Mogiarty has escaped!"

"How?"

"Disguised as a policeman."

"The cunning fiend!"

"But I know where he has gone, Watsup. You have ten minutes to go home and pack. Then we will be taking the cross-channel ferry, assorted trains, a brougham, a dog-shay, an omnibus, and two donkeys. One each."

"But—Soames! Beatrix and I have been married for less than a month! I cannot leave—"

"Your new bride will have to get used to this kind of thing eventually, Watsup, if we are to continue our collaboration."

"True, but—"

"No time like the present. Absence makes the heart grow fonder. A dog is a man's best—well, enough clichés. Her brother will take care of her while you are gone. An absence of six weeks should be ample."

I realised that he would not ask this of me without a compelling reason. He needed me. I must rise to the occasion, no matter the personal cost. "Very well," I said, dire forebodings notwithstanding. "Beatrix will understand. Where are we going?"

"To the Schtickelbach Falls," he said quietly.

I gave an involuntary shudder. It was a name to strike terror into the heart of even the most accomplished mountaineer. "Soames! That is suicide!"

He shrugged. "It is where we shall find Mogiarty. But first we must get there." He pulled out a map.

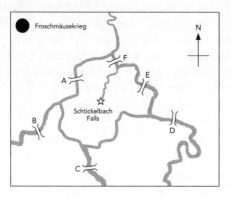

Soames's map

"The map shows the appropriate region of Switzerland. Observe the network of rivers. They rise in the north and flow across the country's borders. The Schtickelbach Falls are at the end of a small river branching off a larger one."

"Where does the river go after the Falls?"

"It plunges beneath the earth, into some underground passage. No one knows where it reappears."

"It is strange geology, Soames."

"The Swiss landscape is a tortured one, Watsup. Now, there are six bridges, which I have marked A, B, C, D, E, F. These are the only bridges within Swiss borders that join the areas of land shown. The omnibus terminus is at the small town of Froschmäusekrieg. From there we will hire donkeys and proceed to the Falls. We must stay in Switzerland: it will be hard enough to cross a national border unnoticed *once*, and it would be foolhardy in the extreme to repeat the attempt. I have already worked out a route, but you may have a better idea."

I studied the map. "Why, it is simple! We cross bridge A."

"No, Watsup. It is too direct. Mogiarty will be expecting that, it is a bridge too far. We must leave bridge A until last in the hope of throwing him off the scent. And we must cross each bridge at most once to minimise the chance of attracting attention and being identified."

"Then we must begin with bridge B," said I. "The only continuation is via C, then D. At that point we have a choice of E or F. Both lead to the Falls, so we may as well use E. Done!"

"As I said, we must leave A until last. Not E."

"Oh, yes. Then we proceed across A—no, that is a dead end with no further connection to the Falls. So we leave A for later and cross F . . . But no: that *also* is a dead end."

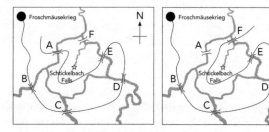

Two paths that do not reach the Falls

Soames grunted non-committally. I checked my analysis. "Perhaps bridge F . . . no. The same problems arise if we use F instead of E, after crossing D. *There is no such path, Soames!*" A thought struck me. "Unless there is a tunnel, or some other way to cross one of the rivers. A ferry? A canoe?"

"There is no tunnel, or ferry, or canoe, and we do not need to cross any rivers. Bridges and dry land suffice."

"Then the thing is impossible, Soames!"

He smiled. "But Watsup: I have already told you that there *is* a route satisfying the stated conditions. Indeed, there are no less than eight essentially different routes—by which I mean that the bridges are crossed in a different order."

"*Eight*? I confess I do not see even one," said I, exasperated.

Is Soames right? See page 303 for the answer.

Is Soames right? See page 303 for the answer.

The Final Problem

From the Memoirs of Dr Watsup

I slept badly and awoke at sunrise, to find Soames already dressed, bright-eyed, and bushy-tailed. "Time for breakfast, Watsup!" he declared in a hearty voice. If he was apprehensive about the forthcoming encounter, he concealed it perfectly.

As soon as we had finished our plates of bread, meat, and Swiss cheese, we mounted our donkeys and made our way up a narrow track. After some miles we tethered our trusty steeds, having reached the base of the Schtickelbach Falls. A wild torrent of water plunged between the sheer sides of a towering chasm, to disappear into a deep hole in the ground, creating a glorious rainbow that glittered in the afternoon sun.

A steep rocky path led to the top of the Falls. As we approached, a silhouette appeared on the skyline above us.

"Mogiarty," said Soames. "There can be no mistaking that evil profile." He took out his pistol and flicked off the safety-catch. "The fiend is trapped, for there is no way down save this pathway. Not any path that a man can take and still live, at any rate. Wait here, Watsup."

"Nay, Soames! I will accomp—"

"You will not. The duty to rid the world of this vile creature is mine alone. I will signal when it is safe for you to join me. Promise me you will remain here until you receive it."

"What signal?"

"You will know when the time comes."

I assented, despite profound misgivings, and he ascended, quickly disappearing from view behind a rocky crag. The last I saw of him was a pair of stout climbing boots.

I waited. All was silence.

Then, suddenly, I heard shouting. The wind carried away the words and I could not make them out. Then I heard the indubitable sounds of a prolonged struggle, and several gunshots. There was a scream, and *something* plunged past me amid the torrent. It was shrouded by spray and moved so rapidly that I could not make it out, but it was roughly the size of a man.

Or two men.

Shocked to the core, I nevertheless did as Soames had told me, and waited.

No signal came.

At last I decided that something had gone wrong, which relieved me of my promise. I clambered up the path. At the top, the rock climbed further skywards in a huge overhang, barring my way. A mossy ledge led towards the precipice from which the waterfall plunged. Of Soames and Mogiarty there was no sign. But, dampened by the spray, the moss held faint traces of footprints.

The prints told a clear tale to anyone who had absorbed lessons in detection from the master. I recognised the indistinct impressions with a chevron pattern as Soames's boots; the other prints with a zigzag were evidently Mogiarty's. The two sets of prints led to the lip of the chasm, and here the ground was churned into thick mud by what had evidently been the struggle I had heard.

I sucked in my breath in horror, for *no footprints returned from that terrible brink.*

I retained enough presence of mind to attempt what Soames would have done, faced with such evidence. Being careful not to superimpose my own footprints—for the local constabulary, inept as they would undoubtedly be, would no doubt wish to inspect the scene—I made a thorough study.

Soames had clearly walked *behind* Mogiarty, for his prints sometimes overlaid those of the criminal, but not *vice versa.* Mogiarty's prints seemed deeper than Soames's, but then, my friend was always light of foot. The dismal conclusion was clear. Soames had pursued Mogiarty to the edge of the precipice; there had been a struggle; both men, no doubt still grappling, had plunged to their doom. Their bodies were now deep underground in some dank chamber, never to be recovered.

I trudged despondently back to the path, where the bare rock showed no footprints. The cliff towered above me, unscalable. I reasoned that had Soames prevailed, he would have signalled and be awaiting my arrival. Had Mogiarty prevailed, he would have been awaiting me instead, armed to the teeth.

There was no conceivable doubt that both men had met the same terrible end.

Yet, even as I began my descent, my friend's voice seemed to echo in my mind, and the tone was mocking. Was my subconscious trying to tell me something? Grief overcame my analytical abilities, and I trudged downhill towards the donkeys. By which I refer to our steeds; however, the Swiss police would be next.

The Return 🔍

From the Memoirs of Dr Watsup

It had been three years since Soames's noble sacrifice had rid the world of Mogiarty. The detective's premises at 222B had passed to his brother Spycraft, and I had taken up medical practice in earnest.

A stooped figure in tattered clothes limped into my surgery. "Is you that doctor bloke? What wrote them dee-tective stories in them magazines?"

I acknowledged my medical status. "I do write, but regrettably the Strand has declined my submissions to date."

"Oh. Must be that other geezer. But you'll do. I gotta terrible pain in me leg, Doc."

"That will be sciatica," I told him. "It is caused by a back problem."

"In me *leg*?"

"The nerves in your leg are trapped somewhere in your spine."

"Oh gawd! I got *nerves* in me leg?"

"Lie on the couch and—" I observed the dirt on his clothes. "No, first let me get a cloth for you to lie on." I turned my back and opened a cupboard.

"No need for that, Watsup," said a familiar voice.

I turned, stared—and fainted.

When I came to, Soames was bending over me waving smelling salts under my nose.

"I do apologise, old chap! I had presumed you had long ago worked out my cunning deception and why it had been necessary."

"Not at all. I thought you dead."

"Ah. Well, you see, when I pushed Mogiarty off the precipice, and saw how the footprints would appear to anyone less astute than myself, I realised in a flash that fate had presented me with a golden opportunity."

"Yes! I see!" I cried. "Although Mogiarty's henchmen in the

British Isles had been apprehended, several remained at large on the continent. If they thought you dead, you could weave your web and trap them. So you supplied some misleading evidence, good enough to convince the bungling Swiss police. Since then you have spent all your waking hours pursuing the criminal scum that remained. One by one you eliminated them. You tracked the last one down in—oh, Casablanca or some other exotic locale—and he will not bother the world again. So now you can reveal that you are still alive."

"A brilliant sequence of deductions, Watsup." I silently congratulated myself. "Though wrong in almost every particular."

Soames explained: "The one thing you got right was that this was a heaven-sent opportunity for me to disappear. But my reasons were not what you imagine. I had run up a significant negative balance betting on the horses, lacked the funds to reimburse my bookmaker, and faced the threat of serious bodily harm. Having finally accumulated the necessary funds, I paid off the debt and rejoined society."

I found it difficult to take in. "I understand your position, Soames. It could happen to the best of us. But how—?"

How did Soames escape and disappear? Make your deductions before reading on, for narrative imperative requires the immediate presentation of the answer.

. . . .

The Final Solution

Soames settled himself beside the fire. "It happened like this, Watsup. When I reached the top of the path, Mogiarty was waiting for me behind a rock. He knocked me out, and was carrying me towards the precipice to hurl me into the abyss. Fortunately I regained consciousness and reached for my pistol. In the ensuing struggle some shots were fired, but no one was hit. Mogiarty's exertions caused him to slip and fall to his death. I was fortunate not to have joined him." He said this in a matter-of-fact voice, as if it had been of little import.

"Intending to summon you, I looked back and saw *a single set of footprints* made by Mogiarty's boots. They led from the path to the brink of the chasm, and nothing led the other way. I saw at once that they were slightly deeper than they should have been for a man of Mogiarty's weight—a clue that I hoped you would spot, Watsup, and knew that the police would not. I then walked *backwards* to the safety of the path, overlaying my own prints, being careful to make them appear as though I had come the other way."

"The thought did cross my mind, Soames. But I dismissed it, since I was unaware of your gambling debts, so I could not think of a motive. But the ledge was empty, the cliff impossible to climb! How did you conceal yourself?"

He ignored my question. "I realised that if I failed to signal, you would eventually decide that your promise was no longer valid, and climb the path. It was but the work of a moment to climb to the ledge above, where the overhang hid me from view. The rest you can guess."

"But—the cliff is unclimbable!" I cried.

He shook his head sadly. "My dear, Watsup, I distinctly remember telling you of a folding grappling-iron that I always carry for such eventualities." (See page 240.) "You really must not forget such vital information. Often one tiny fact is all it takes to unlock a great mystery."

I hung my head, for I had overlooked that item of equipment until that very moment. I attempted a wry grin. "Why, Soames! How abs—olutely, uh, ingenious!"

He gave a thin smile, and changed the subject. "A cup of tea, Watsup?"

"That would be delightful, Soames."

"Then I shall ask Mrs Soapsuds—"

The door swung open and our landlady's head poked around it. "Can I be of service, Mr Soames?"

"—to make us a pot," sighed the detective.

The Mysteries Demystified

or, if not, cast in a new light by sundry extracts from Dr John Watsup's extensive archive of case notes, press cuttings, and Soamesian memorabilia; with occasional contributions from other sources

The Scandal of the Stolen Sovereign

Magnifying glass in hand, Soames inspected every inch of the Glitz's kitchens and accounts. He had the carpets lifted to see what was beneath—a remarkable collection but not of relevance to this tale—and searched Manuel's cramped lodgings in the attic. He sampled the contents of several bottles in the bar. In fact, he had arrived at his conclusions before His Lordship had finished describing the facts of the case, but it would never do to make the process look easy, and the opportunity of a free malt whiskey should not be declined without good reason.

The owner of the Glitz Hotel was waiting in a sumptuously furnished private room, pacing the floor and glaring.

"Have you recovered my stolen sovereign, Soames?"

"No, my Lord."

"Pah! I knew I should have tried Mr Sher—"

"I have not recovered it because there *was* no stolen sovereign. It was never missing in the first place."

"But £27 plus £2 do not make £30!"

"I agree. But there is no reason why they should. The sums add up if you do them correctly." And Soames wrote:

Armstrong	Bennett	Cunningham	Manuel	Glitz Hotel
10	10	10	0	0
0	0	0	0	30
0	0	0	5	25
1	1	1	2	25

"The sum of £30 is no longer the issue," said Soames. "It was, after all, the *wrong bill*. The men have now paid £27, My Lord, and we should *subtract* £2 to get the £25 owed to the hotel. Not add it."

"But—"

"Your original calculation appears to make sense because the numbers 29 and 30 are so close together. But suppose, for example, that the bill had actually been £5, so the waiter was given £25 to return, keeping £1 as a tip and giving them each £8. Now the men had each paid £2, a total of £6. Manuel has retained only £1. The total of these two amounts is £7. You would now ask: where has the other £23 gone? But the actual bill was £5, and the hotel has been paid exactly that. So how can £23 be *missing* from the hotel's share? It has been shared by the three customers, who have given a small part of it to Manuel."

Humphshaw-Smattering turned pink. "Humph," he said. "Pshaw." He pulled himself together. "Your fee, sir?"

"Twenty-nine sovereigns," said Soames, not missing a beat.

Number Curiosity

1001
100001
10000001
1000000001
100000000001
100000000000000001

I also asked why it works. That's a harder question because

you have to think and not just calculate. Instead of a formal proof, let's just look at a typical case: 11×909091. First, rewrite it in the opposite order as 909091×11. This is $909091 \times 10 + 909091 \times 1$; that is, $9090910 + 909091$. Add them like this:

```
9  0  9  0  9  1  0   +
   9  0  9  0  9  1
_____
```

What next? Starting from the right, $0 + 1 = 1$, so we get

```
9  0  9  0  9  1  0   +
   9  0  9  0  9  1
_____
                  1
```

Then $1 + 9 = 0$ carry 1:

```
9  0  9  0  9  1  0   +
   9  0  9  0  9  1
_____
               0  1
            1
```

Now we have to add the carry to the 9 and 0, which again gives 0 carry 1. This leads to a cascade of carries, each converting a 9 to 0 carry 1, until we get to

```
9  0  9  0  9  1  0   +
   9  0  9  0  9  1
_____
0  0  0  0  0  0  1
1
```

Finally only the carry digit is left, leading to the answer

```
   9  0  9  0  9  1  0   +
      9  0  9  0  9  1
_____
1  0  0  0  0  0  0  1
```

Track Position

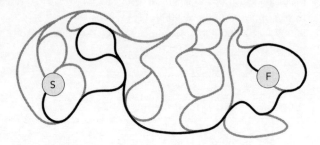

Maze solution

For further information see R. Penrose, Railway mazes, in *A Lifetime of Puzzles*, (eds. E. D. Demaine, M. L. Demaine, T. Rodgers), A.K. Peters, Wellesley MA 2008, 133–148.

Pictures of the Luppitt Millennium Monument can be found at:

http://puzzlemuseum.com/luppitt/lmb02.htm

Soames Meets Watsup

"A decimal point?" Watsup hazarded. "No, you asked for a whole number." He paused, struck by a sudden realisation. "Did you tell me that the symbol must go *between* the two digits, Mr Soames?"

"No."

"Did you insist that the digits be separated by a space?"

"My drawing was perhaps ambiguous, but I did not specify a space."

"I thought so. Would *this* meet your conditions?" And Watsup wrote:

$$\sqrt{49}$$

"Which equals 7."

Geomagic Squares

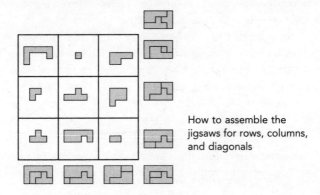

How to assemble the jigsaws for rows, columns, and diagonals

What Shape is an Orange Peel?

Laurent Bartholdi and André Henriques. Orange peels and Fresnel integrals, *Mathematical Intelligencer* 34 No. 4 (2012) 1–3.

You can download a similar article from arxiv.org/abs/ 1202.3033.

How to Win the Lottery?

No. The statements made are all correct, but the deduction is fallacious.

To see why, consider the lottery that runs every week in the little known province of Lilliputia. Here there are only three balls—1, 2, 3—and two of them are drawn. You win by getting those two right.

There are three possible draws:

    12   13   23

and these are all equally likely.

The first number is 1 with probability 2/3, more likely than 2 with probability 1/3 or 3 with probability 0.

The second number is 3 with probability 2/3, more likely than 2 with probability 1/3 or 1 with probability 0.

So, by the same argument, punters should choose 13 to maximise their chances. However, each of the three possibilities is equally likely, so this is rubbish.

In general, 1 is more likely to be the smallest number in the draw because in this case *there are more larger numbers* than there are for any other choice. Not because 1 is more likely to be drawn. The same effect applies to the other positions, but not as obviously.

The Green Socks ~~Caper~~ Incident

"From my deep knowledge of London's low life, it is immediately obvious who the culprit is," Soames announced.

"Who?"

"That is of no concern until we have formal proof of his guilt, Watsup. Nothing less will convince Inspector Roulade of the Metropolitan Police when we present him with our conclusions. First, we must list the possible ways to distribute the colours among the garments."

"I can do that," said Watsup. "I have some small command of elementary combinatorics. It has proved useful when deciding which limb to amputate first." And he wrote:

BGW   BWG   GBW   GWB   WBG   WGB

"The letters denote the colours of the garments, in the order jacket, trousers, socks," Watsup explained. "No colour repeats, because of the witness reports, so the only possibilities are the six permutations of the three letters."

"Very good," said Soames. "And what should our next step be?"

"Uh—to tabulate all of the ways to distribute the garments among the three men. That will take some time, Soames, because there are . . . uh, $6 \times 5 \times 4$ . . . 120 combinations."

"Not so, Watsup. With a little thought we can eliminate most of them at the outset. Let us begin by focusing on just one of the suspects—George Green, say. Suppose, for the sake of argument, that Green wears a green jacket, brown trousers, and white socks: case GBW."

"Ah, but does he?"

"So I hypothesise, for the sake of argument. If that is correct, it follows that the other two suspects cannot wear a green jacket, or brown trousers, or white socks, since only one of each type of garment has a given colour. So for those men we can eliminate GWB, BGW, and WBG from the five possibilities remaining. That leaves only BWG and WGB. Which, you observe, are cyclic permutations of GBW. We can assign these choices to Bill Brown and Wally White in only two ways." Soames began to compile his table:

	George Green	Bill Brown	Wally White
1.	GBW	BWG	WGB
2.	GBW	WGB	BWG

"But Soames," cried Watsup, "perhaps George Green does not wear the garments GBW!"

"Quite possibly," said Soames, unperturbed. "These are merely the top two rows of my table. I can make similar deductions for the other five possible lists for George Green. And, of course, again the permutations are cyclic. There are thus twelve possibilities altogether."

Watsup copied out the resulting table:

	George Green	Bill Brown	Wally White
1.	GBW	BWG	WGB
2.	GBW	WGB	BWG
3.	GWB	WBG	BGW
4.	GWB	BGW	WBG
5.	BGW	GWB	WBG
6.	BGW	WBG	GWB
7.	BWG	WGB	GBW
8.	BWG	GBW	WBG
9.	WGB	GBW	BWG
10.	WGB	BWG	GBW

| 11. | WBG | BGW | GWB |
| 12. | WBG | GWB | BGW |

When he had finished, Soames nodded. "And now, my dear Watsup, all that remains is to use the evidence to eliminate the impossible combinations—"

"Because then, whatever remains, however improbable, must be true!" Watsup cried.

"I could not have put it better myself. Though in this case, the most improbable feature is that only one of these villains was involved. I would have expected a conspiracy.

"Anyway, Constable Wuggins—an admirable fellow, Watsup, who makes up in perseverance for what he lacks in imagination—stated that Brown's socks were the same colour as White's jacket. That means that Brown's triple of letters must end with the same letter that begins White's triple. That eliminates rows 1, 3, 5, 7, 9, 11, and reduces the table to

	George Green	**Bill Brown**	**Wally White**
2.	GBW	WGB	BWG
4.	GWB	BGW	WBG
6.	BGW	WBG	GWB
8.	BWG	GBW	WGB
10.	WGB	BWG	GBW
12.	WBG	GWB	BGW

"Next, I determine which combinations satisfy the good Constable's second condition: that the person whose name was the colour of White's trousers wore socks whose colour was not the name of the person wearing a white jacket. This is purely a matter of keeping a clear head. For example, in row 4 White's trousers are brown, so the person whose name is the colour of White's trousers is Brown. His socks are white. Is White's jacket a different colour from white? No: it *is* white. So we delete row 4."

"I'm not sure I quite—"

"Oh, very well, let me draw up another table!" And Soames wrote:

	colour of White's trousers	corresponding person	colour of his socks	person with white jacket	same?
2.	W	W	G	B	yes
4.	B	B	W	W	no
6.	W	W	B	B	no
8.	G	G	G	W	yes
10.	B	B	G	G	no
12.	G	G	G	G	no

"Only rows 2, and 8 remain. Which further reduces the table to

	George Green	Bill Brown	Wally White
2.	GBW	WGB	BWG
8.	BWG	GWB	WGB

"Finally, Constable Wuggins tells us that the colour of the jacket of the person whose name is the colour of Green's socks is different from the colour of Brown's trousers."

	colour of Green's socks	corresponding person	colour of his jacket	colour of Brown's trousers	different?
2.	W	W	B	G	yes
8.	G	G	B	B	no

"That eliminates row 8, leaving only row 2.

"So now it remains only to see who was wearing the green socks in row 2. As I suspected from the start, it was Wally White with the entry BWG."

Consecutive Cubes

$$23^3 + 24^3 + 25^3 = 12,167 + 13,824 + 15,625 = 41,616 = 204^2$$

This can be found by trying numbers in turn. A more systematic method is to let the middle number be n and observe that $(n-1)^3 + n^3 + (n+1)^3 = 3n^3 + 6n = m^2$ for some m. So $m^2 = 3n(n^2 + 2)$. The terms 3, n, $n^2 + 2$ have no common factors aside from perhaps 2 and 3. Therefore any prime factor greater than 3 must occur to an even power (perhaps 0) in both n and $n^2 + 2$. The first two numbers to survive this test are 4 and 24, and 24 provides a solution but 4 does not.

Adonis Asteroid Mousterian

The numbers should be assigned like this:

Order 3: A = 0, D = 3, I = 2, N = 0, O = 1, S = 6.

Order 4: A = 0, D = 12, E = 1, I = 2, O = 3, R = 8, S = 0, T = 4.

Order 5: A = 0, E = 1, I = 2, M = 0, N = 5, O = 3, R = 10, S = 15, T = 20, U = 4.

The squares become

3	2	7
8	4	0
1	6	5

ADONIS

0	10	13	7
15	5	2	8
6	12	11	1
9	3	4	14

ASTEROID

6	0	12	18	24
17	23	9	1	10
4	11	15	22	8
20	7	3	14	16
13	19	21	5	2

MOUSTERIAN

Changing letters to numbers and adding

For more magic word squares and similar constructions, see: Jeremiah Farrell, Magic square magic, *Word Ways* 33 (2012) 83–92. Available at:

http://digitalcommons.butler.edu/wordways/vol33/iss2/2

Two Square Quickies

1   923187456, the square of 30384.

Since we want the largest number of its type, it's a fair bet that the answer starts with 9, so this really has to be tried first, even if it turns out to be false. So it must lie between 912345678 and 987654321, bearing in mind that all digits are different and there is no 0. The square roots of these are 30205.06 and 31426.96. So all we have to do is square the numbers between 30206 and 31426 and see which give all nine nonzero digits. There are 1221 such numbers. Working backwards from 31426 we eventually get to 30384. Now that we've found a solution starting with 9, we don't need to worry about starting with 8 or lower.

2   139854276, the square of 11826.

The way to find this is similar.

The Adventure of the Cardboard Boxes

1   The boxes have dimensions $6 \times 6 \times 1$ and $9 \times 2 \times 2$.

Suppose the dimensions of the boxes are x, y, z and X, Y, Z. Their volumes are xyz and XYZ. The length of ribbon is $4(x + y + z)$ and $4(X + Y + Z)$. Dividing out the factor of 4, we must solve

$$xyz = XYZ$$
$$x + y + z = X + Y + Z$$

in nonzero whole numbers. That is, find two triples of numbers with the same product and the same sum. The smallest solution is $(x, y, z) = (6, 6, 1)$ and $(X, Y, Z) = (9, 2, 2)$. The product is 36 and the sum is 13.

2   The smallest solution for three boxes is (20, 15, 4), (24, 10, 5), and (25, 8, 6). Now the product is 1,200 and the sum is 39.

In passing, we can answer a third question, which did not feature in Soames's investigation:

3   Suppose that the ribbons are tied in the manner of the

left-hand picture, with x being width, y depth, and z height. Then the equations become

$$xyz = XYZ$$
$$x + y + 2z = X + Y + 2Z$$

If we replace x, y, z by x, y, $2z$, and similarly for X, Y, and Z, we are again seeking two triples of numbers with the same product (now $2xyz = 2XYZ$) and sum. However, z and Z must be *even*. This is the case in solution (1) if we arrange the sides in the right order, and it leads to the smallest solution (6, 1, 3) and (9, 2, 1).

My attention was drawn to this problem by Moloy De from Kolkata, India, who has also found the smallest sets of four, five, and six whole numbers with the same sum and product:

Four packages

> (54, 50, 14) (63, 40, 15) (70, 30, 18) (72, 25, 21)
> sum = 118, product = 37,800.

Five packages

> (90, 84, 11) (110, 63, 12) (126, 44, 15) (132, 35, 18)
> (135, 28, 22)
> sum = 185, product = 83,160.

Six packages

> (196, 180, 24) (245, 128, 27) (252, 120, 28) (270, 98, 32)
> (280, 84, 36) (288, 70, 42)
> sum = 400, product = 846,720.

The RATS Sequence

The next term is 1345.

The rule is: 'Reverse, Add, Then Sort'. RATS. By 'sort' I mean rearrange into ascending order. Any zeros are omitted. For example:

> $16 + 61 = 77$ already in numerical order
> $77 + 77 = 154$, reorder as 145
> $145 + 541 = 686$, rearrange as 668
> $668 + 866 = 1534$, rearrange as 1345

John Horton Conway has conjectured that whichever number you start with, eventually the sequence either goes round and round some repeating cycle or gets into the ever-increasing sequence

$$123^n 4444 \rightarrow 556^n 7777 \rightarrow 123^{n+1} 4444 \rightarrow 556^{n+1} 7777 \rightarrow \cdots$$

where the n indicates not a power but n identical digits repeated.

Mathematical Dates

The next triple palindrome day will be on 21:12 21/12 2112. The next palindrome was 20:02 30/03 2002 (British system).

The Hound of the Basketballs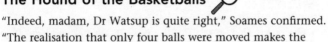

"Indeed, madam, Dr Watsup is quite right," Soames confirmed. "The realisation that only four balls were moved makes the required arrangement obvious."

"What is it?"

"That, madam, is information which, according to your own statements, should only be disclosed to the senior extant member of the male line."

"Namely, Lord Edmund Basquet," I pointed out. "Who is in a coma. Which poses a considerable diff—"

"Rubbish!" said Lady Hyacinth. "You can tell me." It was evident from her face that nothing would deter her from this course of action.

"Very well," said Soames, sketching rapidly. "The pood—er, giant slavering hound—must have moved the four basketballs shown in white to the positions shown in black. Or either of the two rotations of this solution. But you said that the orientation of the configuration does not matter."

Now I understood the point of his obscure query earlier.

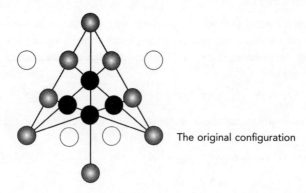

The original configuration

"Wonderful!" said Lady Hyacinth. "I will instruct Willikins to make it so."

"But will that not infringe the conditions of the ceremony?" I enquired.

"Of course, Dr Watsup. But there is no rational reason to fear any adverse consequences. That ancient taboo is little more than a load of old, er, superstition."

A month later, Soames handed me the *Manchester Garble*.

"Great Scott!" I cried. "Lord Basquet has died and Basquet Hall has burned to the ground! The family's insurance company has declined payment because Malevolent Forces of Pure Evil are excluded, and the family is now ruined! Lady Hyacinth has been confined to an asylum for the incurably insane!"

Soames nodded. "Pure coincidence, I'm sure," he said. "Perhaps with hindsight I should have told Her Ladyship about the poodle."

Digital Cubes

370, 371, and 407.

Despite this problem allegedly having no mathematical significance, you have to be quite good at maths to find the four solutions, and even better to prove there aren't any others.

I'll sketch one approach.

Since numbers with leading zeros are excluded, there are only 900 combinations to try. But we can cut this down. The cubes of the ten digits are 0, 1, 8, 27, 64, 125, 216, 343, 512, and 729. The sum of the three digits is 999 or less, so we can rule out any number containing two 9's, two 8's, an 8 and a 9, and so on.

Suppose one digit is 0. Then the number is a sum of two cubes from the list. Of the 55 such pairs only $343 + 27 = 370$ and $64 + 343 = 407$ have the required property.

We may now assume no digit is 0. Suppose one of them is 1. A similar calculation leads to $125 + 27 + 1 = 153$ and $343 + 27 + 1 = 371$.

We may now assume no digit is 0 or 1. Now we have a smaller list of cubes to work with. And so on.

Short cuts, such as considering odd and even numbers, shorten the calculations further. It's a bit long, but a systematic approach (as Soames always recommends) gets there without any serious problems along the way.

Narcissistic Numbers

Here we allow leading zeros.

Fourth powers: 0000   0001   1634   8208   9474
Fifth powers: 00000   00001   04150   04151   54748   92727   93084

Clueless! 🔍

3	2	1	4
2	1	4	3
1	4	3	2
4	3	2	1

Watsup's solution to clueless pseudoku

"Soames!" I cried. "I've solved it!"

"Yes, the murderess was Gräfin Liselotte von Finkelstein, riding her thoroughbred *Prinz Igor* and towing three carthorses behind to obscure the tracks in the—"

"No, no, Soames, not your case! The puzzle!"

He gave my scrawled solution a cursory glance. "Correct. A lucky guess, no doubt."

"No, Soames, I reasoned it out using the logical principles that you have impressed upon my consciousness. First, I realised that the numbers in each region must sum to 20."

"Because the total of the numbers in all squares is $(1+2+3+4) \times 4 = 40$, to be divided equally between the two regions," Soames said dismissively.

"Exactly. Now, once it had occurred to me to concentrate on the *larger* region, the solution began to fall into place. That region has four cells in a row along the bottom, which must contain 1, 2, 3, 4 in some order, and those add to 10, whatever the order may be. So the remaining three rows must also add to 10. The only way this can happen is if the top row contains 1, 2, 3 in some order; the second row contains 1, 2 in some order, and the third row contains 1."

"Why?"

"Any other choice would make the total too big."

"You are indeed learning, Watsup. Very good: continue."

I smiled at this faint praise, since getting *any* praise from Soames was like making marmalade from the Isle of Wight. "Well—it's now easy to verify that there is only one possible way to complete the arrangement. The numbers in the second piece are forced: for example, the top row must end in 4, and then the other 4's must go down the diagonal; then the two 3's are forced, and finally the 2 goes in the remaining position."

This puzzle was invented by Gerard Butters, Frederick Henle, James Henle, and Colleen McGaughey, Creating clueless puzzles, *The Mathematical Intelligencer* 33 No. 3 (Fall 2011) 102–105. See also the website:

http://www.math.smith.edu/~jhenle/clueless/

A Brief History of Sudoku

The two basically different solutions to Ozanam's puzzle are:

A♠	K♥	Q♦	J♣	A♠	K♥	Q♦	J♣
Q♣	J♦	A♥	K♠	J♦	Q♣	K♠	A♥
J♥	Q♠	K♣	A♦	K♣	A♦	J♥	Q♠
K♦	A♣	J♠	Q♥	Q♥	J♠	A♣	K♦

Remember: each of these gives rise to 576 solutions by permuting the values and suits, so don't be surprised if your solutions look different. If you start with the top row A♠ K♥ Q♦ J♣ (or rearrange your solution into this form) you need only think about permuting the other three rows.

Once, Twice, Thrice

2	1	9	2	7	3	3	2	7
4	3	8	5	4	6	6	5	4
6	5	7	8	1	9	9	8	1

The Case of the Face-Down Aces

"It is all trickery, Watsup. With the right preparation, the trick works automatically, no matter which sequence of folds the audience chooses."

"Dashed clever, what?" said I.

Soames grunted. "When Whodunni prepared the pack, he placed the four aces in positions 1, 6, 11, and 16 from the top down. So, when the pack was dealt into a square, the aces lay along the diagonal from top left to bottom right. But they were face down, so of course you would not have been aware of the deception.

"Imagine turning the cards that lie along the diagonal face up. Then the square would have a pattern like a chessboard, with the aces along the diagonal:

Whodunni's initial arrangement with the diagonal cards turned over

"Now, this arrangement has a wonderful mathematical property. *However* you fold the square, at any stage the cards that end up in a given position will all face the same way: either all up or all down."

"Really?"

"Let us try. For instance, we might begin by folding along the central vertical line. Think of the top row of cards. The third card (up) turns over (down) and is placed on top of the second card— also down. And the fourth card (down) turns over (up), and is placed on top of the first card—also up."

I began dimly to see how it worked. "And the same goes for the other rows?"

"Indeed. This first fold creates a rectangle, made from cards or small piles of cards. The cards in each pile all face the same way (up or down), and the set of piles has the same chessboard pattern of ups and downs as the original set of cards. So the same thing happens for the next fold, and the next. By the time we reach a single pile, all of the cards in the pile face the same way."

"Yes, but—when we started, the cards on the diagonal were the wrong way up compared to the chessboard pattern," said I.

It was intended as an objection, but he beamed at my insight. "Exactly! So, after folding, they will *again* be the wrong way up. So instead of a pile of sixteen cards all facing the same way, you will have a pile with twelve cards facing one way, and the four aces facing the other."

It was devilishly cunning.

The chessboard pattern has what mathematicians call 'colour symmetry'. The fold lines act like mirrors, and the mirror image of each card sits on top of a card that faces the other way. This idea is used to study how the atoms in crystals are arranged. The cunning bit is to turn the mathematics into an effective card trick. Whodunni didn't do that. Following his usual *modus operandi*, he stole the trick from its inventor Arthur Benjamin, a mathematician and magician at Harvey Mudd College in California.

Jigsaw Paradox

Neither shape is a triangle. The first bulges slightly upwards along the sloping edge, the second bulges slightly downwards. That's where the missing square has gone.

The Catflap of Fear

Soames nodded in satisfaction. "I have it, Watsup! Cirrhosis goes out, Dysplasia goes out, Aneurysm goes out, Cirrhosis comes back in, Borborygmus goes out, Cirrhosis goes out."

We began the delicate process of enticing cats out through the flap and stuffing them back in. "Take care, Soames!" I whispered. "One mistake and this entire area will be a smoking crater. I do not wish to present myself or my cats at the Pearly Gates quite yet. I am wearing unpressed trousers, and the cats need brushing."

"Do not concern yourself, Watsup," said he, grabbing Cirrhosis before the wretched animal could hoof it over the fence. "You can have total confidence in my solution."

"I do not doubt it, Soames," I replied, casting hastily about for something solid to hide behind. "Er—how did you make your deductions?"

He borrowed my notebook and a pencil.

"There are 16 possibilities for which cats are in the house: ABCD, ABC, ABD, and so on until none are present (call this *). Use an arrow → to denote a possible move: one cat through the flap.

"The first condition rules out AC and ABC. The second rules out BD and BCD. The third rules out AD. The fourth rules out CD. The fifth rules out the change A → *. The sixth rules out B → *.

"Now, ABCD → ACD or ABD. However, ACD → AC, AD, or CD, all of which are ruled out. Therefore ABC → ABD. Since ABD → AD and ABD → BD are ruled out, we must have ABD → AB. But AB → A is pointless since A can't go out if no others are present. So AB → B. However, B cannot then go out, so some other cat must come back in. But B → AB involves A coming straight back in, and B → BD is ruled out, so B → BC. Now BC → C → *.

"There is also a visual way to see this, which in some ways is simpler," he added, and sketched a diagram. "This picture shows all 16 possible combinations of cats, with the thin lines representing possible changes as a cat goes in or out. The black dots are ruled out, the two crosses rule out two of the lines. The bold line is the *only* path that runs from ABCD to * using only permitted dots and lines, but never backtracks."

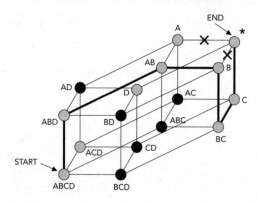

Catflap conditions

Shortly afterwards, I was reunited with my furry friends. "Soames, how can I ever thank you?" I cried, clasping the animals joyously to my chest.

He looked down at his jacket. "By brushing your cats more often, Watsup."

Pancake Numbers

1   No.

2   Some stacks of four pancakes need four flips; for example, the one below. See the figure below that for the other two. No stack needs more than four flips.

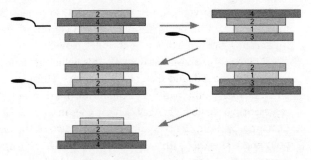

A stack that needs four flips

Here's a systematic method for proving those statements. The diagram shows the required final arrangement 1234 at the top, where the sizes are listed in order from the top down. We work backwards from this. The second line shows all arrangements that can be obtained from 1234 by one flip. These are *also* the arrangements that can reach 1234 in one flip. (The same flip done twice puts everything back where it was.) The third line shows all arrangements that can be obtained from the first line by one flip. These are also the arrangements that can reach 1234 in two flips. Notice that precisely one entry in the third row can be reached from two in the second, namely 1324. So the structure of the diagram looks slightly asymmetric there.

Rows 1, 2, 3 contain 21 of the 24 possible stack orders. The missing ones are 2413, 3142, and 4231. Row 4 shows how these can be obtained from row 3 by one more flip—or, reversing the series of flips, how to convert them to 1234 in four flips. (The other links to row 4 are omitted since they make the diagram more complicated and we don't need them.) The figure above in answer 2 is the arrangement 2413 converted into stacks.

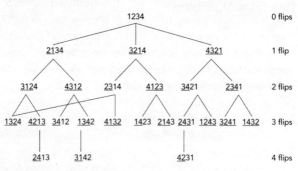

Stacks that need 1, 2, 3, or 4 flips to put them in order

3   Either the biggest pancake is at the top, or it's not. If not, insert the spatula under it and flip. Now it's at the top. Insert the spatula at the bottom of the stack and flip the entire thing. Now the biggest is at the bottom. So at most two flips will put the biggest pancake at the bottom. Leave it there and repeat for the next biggest pancake: at most two flips will place it directly on top of the biggest. Repeat for the next biggest, and so on. It takes at most two flips to get each successive pancake in its correct position, so at most $2n$ flips will achieve this for the entire stack of n pancakes.

4   $P_1 = 0$, $P_2 = 1$, $P_3 = 3$, $P_4 = 4$, $P_5 = 5$.

The pancake-sorting problem goes back to Jacob Goodman in 1975, who published it under the pseudonym Harry Dweighter. (Say it aloud and pronounce the surname as 'Dwayter' to get the joke.) The solution is known for all n up to 19, but not for 20. The results are:

n	1	2	3	4	5	6	7	8	9	10
P_n	0	1	3	4	5	6	8	9	10	11

n	11	12	13	14	15	16	17	18	19	20
P_n	13	14	15	16	17	18	19	20	22	?

The pancake numbers tend to run in sequence, going up by 1 as n increases. For example P_n is 3, 4, 5, 6 when $n = 3, 4, 5, 6$. But this pattern goes wrong when $n = 7$ because $P_7 = 8$, not 7. After that, there is a jump of 2 at $n = 11$, and again at $n = 19$.

The upper estimate of $2n$ flips, my answer to question 3, can be improved. In 1975 William Gates (yes, *the* Bill Gates) and Christos Papadimitriou replaced it by $(5n + 5)/3$.

Gates and Papadimitriou also discussed the *burnt pancake problem*. Here each pancake is burnt on one side, which could be the top or the bottom, and you have to get all the burnt sides on the bottom as well as stacking the pancakes in the right order. In 1995 David Cohen proved that the burnt pancake problem needs at least $3n/2$ flips, and can be solved with at most $2n-2$ flips.

If you're thinking of tackling $n = 20$, bear in mind that there are

$$2{,}432{,}902{,}008{,}176{,}640{,}000$$

different stacks to start from.

The Case of the Cryptic Cartwheel

"The diameter of the wheel is 58 inches, of course," said Soames. "It is an elementary application of Pythagoras's theorem."

I thought about this. I have some small facility in geometry and algebra. "Let me try, Soames. I take the radius of the wheel to be r. The shaded triangle in your diagram is right-angled, with hypotenuse r and the other sides being $r-8$ and $r-9$. So, as you hinted, I can apply Pythagoras, getting

$$(r - 8)^2 + (r - 9)^2 = r^2$$

That is,

$$r^2 - 34r + 145 = 0$$

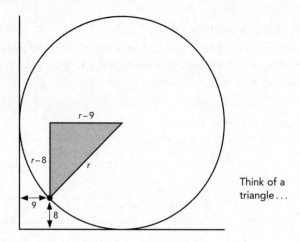

Think of a triangle...

I stared at the symbols, temporarily halted.

"The quadratic factorises, Watsup:

$$(r - 29)(r - 5) = 0$$

"So it does! Which means that the solutions are $r = 29$ and $r = 5$."

"Yes. But you must remember that the diameter is $2r$, which is 58 or 10. However, the solution 10 inches is ruled out because the diameter is more than 20 inches. All that remains—"

"Is 58 inches," I finished for him.

The V-shaped Goose Mystery

Florian Muijres and Michael Dickinson, Bird flight: Fly with a little flap from your friends, *Nature* 505 (16 January 2014) 295–296.

Steven J. Portugal and others, Upwash exploitation and downwash avoidance by flap phasing in ibis formation flight, *Nature* 505 (16 January 2014) 399–402.

Amazing Squares

The main idea can be expressed in full generality using algebra, but I'll forego formalities and illustrate it by example. Look at the process in reverse, starting with

$$9^2 + 5^2 + 4^2 = 8^2 + 3^2 + 7^2$$

and expanding it to

$$89^2 + 45^2 + 64^2 = 68^2 + 43^2 + 87^2$$

It's easy to check the first equation, which is how everything gets started, but why does the *second* equation hold?

The actual value of a two-digit number [*ab*] is $10a + b$. So we can write the left-hand side as

$$(10 \times 8 + 9)^2 + (10 \times 4 + 5)^2 + (10 \times 6 + 4)^2$$

which is

$$100(8^2 + 4^2 + 6^2) + 20(8 \times 9 + 4 \times 5 + 6 \times 4) + 9^2 + 5^2 + 4^2$$

Similarly, the right-hand side becomes

$$100(6^2 + 4^2 + 8^2) + 20(6 \times 8 + 4 \times 3 + 8 \times 7) + 8^2 + 3^2 + 7^2$$

Comparing these, the first terms are equal because $6^2 + 4^2 + 8^2$ is just $8^2 + 4^2 + 6^2$ in a different order, and the third terms are equal because we started from those. So we just need to see whether the middle terms are equal; that is, whether

$$8 \times 9 + 4 \times 5 + 6 \times 4 = 6 \times 8 + 4 \times 3 + 8 \times 7$$

In fact, both are 116.

Everything up to this point would have worked if we'd used any three single-digit numbers in place of 8, 4, 6. So we just have to choose these numbers to make the final expressions equal.

The rest of the explanation is similar.

The Thirty-Seven Mystery 🔍

With some prodding from Soames along the way, I eventually realised that the key to the mystery is the equation $111 = 3 \times 37$. The three-digit numbers that produce long lists of repeated digits when subjected to my procedure turn out to be those that are multiples of 3. This is the case for 123, 234, 345, 456, and 126, for example. For such numbers the procedure is equivalent to multiplying many repetitions of a smaller number, one third the size, by 3×37, which is 111.

As an example, consider Soames's 486. This is 3×162. Therefore, multiplying 486486486486486486 by 37 is the same as multiplying 162162162162162162 by 111. Since $111 = 100 + 10 + 1$ we can do this by adding together the numbers

> 16216216216216216200
> 1621621621621621620
> 162162162162162162

From right to left, we obtain $0 + 0 + 2 = 2$, then $0 + 2 + 6 = 8$. After that, we get $2 + 6 + 1$, $6 + 1 + 2$, $1 + 2 + 6$, over and over again, until we get neat the left-hand end. But these are the same three numbers added in various orders, so the result is the same in each case—namely 9.

When Soames first explained this to me, I raised an objection. "Yes, but what if the three numbers add to more than 9? Then there is a carry digit!"

His reply was brief and to the point. "Yes, Watsup: the *same* carry digit every time." I eventually realised that this means that again, one digit will repeat many times.

"There are, of course, more formal proofs," Soames remarked, "but I think this one makes the idea clear." After which he returned to his chair with a pile of newspapers and said no more that evening, while I went downstairs to beg a plate of Gorgonzola sandwiches from Mrs Soapsuds.

[This item was inspired by some observations made by Stephen Gledhill.]

Average Speed

We're using the wrong mean. We should be using the harmonic mean (explained below), not the arithmetic mean.

We normally define the 'average speed' for some journey to be the total distance travelled divided by the total time taken. If the journey is broken up into pieces, then the average speed for the whole trip is not, in general, the arithmetic mean of the average speeds for the pieces. If the pieces are travelled in equal *times*, the arithmetic mean works, but it doesn't if they cover equal *distances*, which is the case here.

Equal times first. Suppose a car drives at speed a for time t, and then at speed b for the same time t. The total distance is $at + bt$, travelled in time $2t$. So the average speed is $(at + bt)/2t$, which equals $(a + b)/2$, the arithmetic mean.

Next, equal distances. Now the car drives distance d at speed a, taking time r. Then it drives distance d again, at speed b, taking time s. The total distance is $2d$, and the total time is $r + s$. To express this in terms of the speeds a and b, observe that $d = ar = bs$. So $r = d/a$ and $s = d/b$. The average speed is therefore:

$$\frac{2d}{\dfrac{d}{a} + \dfrac{d}{b}}$$

This simplifies to $2ab/(a + b)$, which is the harmonic mean of a and b. It is the reciprocal of the arithmetic mean of the reciprocals of a and b, where the reciprocal of x is $1/x$. This occurs because the time taken is proportional to the reciprocal of the speed.

Four Clueless Pseudoku

total 15

1	3	4	2	5
4	1	2	5	3
2	5	1	3	4
5	4	3	1	2
3	2	5	4	1

total 14

2	1	3	4	6	5
1	5	6	3	4	2
4	6	2	5	3	1
3	4	5	2	1	6
6	2	4	1	5	3
5	3	1	6	2	4

total 14

5	3	1	4	2	6
1	5	3	6	4	2
3	1	5	2	6	4
2	6	4	5	3	1
4	2	6	1	5	3
6	4	2	3	1	5

total 25

1	2	4	5	3
3	1	2	4	5
5	3	1	2	4
4	5	3	1	2
2	4	5	3	1

or flip in
main
diagonal

Clueless pseudoku answers

These puzzles also come from Gerard Butters, Frederick Henle, James Henle, and Colleen McGaughey. Creating clueless puzzles, *The Mathematical Intelligencer* 33 No. 3 (Fall 2011) 102–105.

The Puzzle of the Purloined Papers

"Charlesworth was the thief," said Soames.

"Are you certain, Hemlock? Much hangs upon your being right."

"There can be no doubt, Spycraft. Their statements are:

Arbuthnot: Burlington did it.

Burlington: Arbuthnot is lying.

Charlesworth: It was not I.

Dashingham: Arbuthnot did it.

We know that one of the men speaks truly and the other three lie. There are four possibilities. Let us try them in turn.

"If only Arbuthnot is telling the truth then his statement

informs us that Burlington is the guilty party. However, Charlesworth is lying, so it was Charlesworth. This is a logical contradiction, so Arbuthnot is not telling the truth.

"If only Burlington is telling the truth then—"

"Charlesworth is lying!" I cried. "So it *was* Charlesworth!"

Soames glared at me for stealing his thunder. "That is so, Watsup, and the other statements are consistent with it. So we already know that Charlesworth is the thief. However, it is worth checking the other two possibilities to avoid even the remote possibility of error."

"Absolutely, old chap," said I.

He took out his pipe but did not light it. "If only Charlesworth is telling the truth then Burlington's statement is false, so Arbuthnot is telling the truth, again a contradiction since he is lying.

"If only Dashingham is telling the truth then the same contradiction arises.

"So the only possibility is that Burlington is the sole person telling the truth, and that confirms that the thief is Charlesworth. As Watsup so astutely deduced."

"Thank you, gentlemen," said Spycraft. "I knew I could rely on you." He gestured and a shadowy figure entered the room. They held a whispered conversation, and he left again. "The Captain's residence will be searched forthwith," said Spycraft. "I am confident the document will be found there."

"Then we have saved the Empire!" I replied.

"Until the next time someone leaves secret documents on the seat of a cab," said Soames drily.

On our way out I whispered to my companion: "Soames, if Spycraft is an expert in prime numbers, what on earth is he doing working in counter-espionage? There can be no possible connection, can there?"

He stared at me for a moment, and shook his head. Whether to confirm the absence of any connection, or to warn me not to pursue the matter, I do not know.

Another Number Curiosity

$$123456 \times 8 + 6 = 987654$$
$$1234567 \times 8 + 7 = 9876543$$
$$12345678 \times 8 + 8 = 98765432$$
$$123456789 \times 8 + 9 = 987654321$$

It's not totally clear what 'ought' to come next: perhaps

$$1234567890 \times 8 + 10$$

which is 9876543130, so the pattern stops. But maybe I should have used $(123456789) \times 10 + 10 = 1234567900$. Now

$$1234567900 \times 8 + 10 = 9876543210$$

Next, $(12345678900) \times 10 + 11 = 123456789011$, leading to

$$12345679011 \times 8 + 11 = 98765432099$$

and so on. If you experiment, a different pattern emerges, which continues indefinitely.

Progress on Prime Gaps

The Elliott-Halberstam Conjecture [Peter Elliott and Heini Halberstam, A conjecture in prime number theory, *Symposia Mathematica* 4 (1968) 59-72] is very technical. Write $\pi(x)$ for the number of primes less than or equal to x. For any positive integer q and a having no factor (other than 1) in common with q, let $\pi(x;q,a)$ be the number of primes less than or equal to x that are congruent to $a \pmod{q}$. This is approximately equal to $\pi(x)/\phi(q)$ where ϕ is Euler's totient function, the number of integers between 1 and $q-1$ that have no factor in common with q. Consider the largest error

$$\max_a \left| \pi(x;q,a) - \frac{\pi(x)}{\phi(q)} \right|$$

The Elliott-Halberstam Conjecture tells us how big this error is: it states that for all $\theta < 1$ and $A > 0$ there exists a constant $C > 0$ such that

$$\sum_{1 \leq q \leq x^\theta} E(x;q) \leq \frac{Cx}{\log^A x}$$

for all $x > 2$. It is known to be true for $\theta < \frac{1}{2}$.

The Sign of One: Part Two

Here's one solution:

$7 =$

$\lceil\sqrt{}\sqrt{}\sqrt{}((\lfloor\sqrt{}\sqrt{}\sqrt{}((\lceil\sqrt{}\sqrt{}\sqrt{}\sqrt{}\sqrt{}((\lfloor\sqrt{}\sqrt{}\sqrt{}((\lceil\sqrt{}\sqrt{}((\lfloor\sqrt{}\sqrt{}(11!)\rfloor)!)\rceil)!)\rfloor)!)\rceil)!)\rfloor)!)\rceil$

See The Sign of One: Part Three, page 115, for an explanation.

Euclid's Doodle

You *could* do it by hand using prime factors, given a day or two. You'd have to work out that

$$44{,}758{,}272{,}401 = 17 \times 17{,}683 \times 148{,}891$$
$$13{,}164{,}197{,}765 = 5 \times 17{,}683 \times 148{,}891$$

Then you'd conclude that the hcf is $17{,}683 \times 148{,}891$, which equals 2,632,839,553.

Using Euclid's algorithm, the whole calculation goes like this:

$(13{,}164{,}197{,}765; 44{,}758{,}272{,}401) \rightarrow (13{,}164{,}197{,}765; 31{,}594{,}074{,}636)$

$\rightarrow (13{,}164{,}197{,}765; 18{,}429{,}876{,}871) \rightarrow (5{,}265{,}679{,}106, 13{,}164{,}197{,}765)$

$\rightarrow (5{,}265{,}679{,}106; 7{,}898{,}518{,}659) \rightarrow (2{,}632{,}839{,}553; 5{,}265{,}679{,}106)$

$\rightarrow (2{,}632{,}839{,}553; 2{,}632{,}839{,}553) \rightarrow (0; 2{,}632{,}839{,}553)$

So the hcf is 2,632,839,553.

123456789 Times X

$$123456789 \times 1 = 123456789$$
$$123456789 \times 2 = 246913578$$
$$123456789 \times 3 = 370370367$$
$$123456789 \times 4 = 493827156$$
$$123456789 \times 5 = 617283945$$
$$123456789 \times 6 = 740740734$$
$$123456789 \times 7 = 864197523$$
$$123456789 \times 8 = 987654312$$
$$123456789 \times 9 = 1111111101$$

These multiples have all nine nonzero digits in some order, *except* when we multiply by something divisible by 3 (that is, 3, 6, and 9).

The Sign of One: Part Three 🔍

Since

$$62 = 7 \times 9 - 1 = 7/.\dot{1} - 1$$

we can use the representation of 7 with two 1's from page 279 to get 62 with four 1's.

For a long time Soames and Watsup despaired of getting 138 with four 1's, but by using Watsup's insight about square roots and factorials, and being systematic, they eventually discovered that it is possible to get 138 using only *three* 1's. Again the starting point is 7 written with two 1's, and then

$$70 = \lfloor \sqrt{7!} \rfloor$$
$$37 = \lceil \sqrt{\sqrt{\sqrt{\sqrt{\sqrt{\sqrt{70!}}}}}} \rceil$$
$$23 = \lceil \sqrt{\sqrt{\sqrt{\sqrt{\sqrt{37!}}}}} \rceil$$
$$26 = \lceil \sqrt{\sqrt{\sqrt{\sqrt{23!}}}} \rceil$$
$$46 = \lfloor \sqrt{\sqrt{\sqrt{\sqrt{26!}}}} \rfloor$$

and finally

$$138 = 46/\sqrt{.\dot{1}}$$

which is a clever way to multiply by 3 using just one extra 1.

Tossing a Fair Coin Isn't Fair

Persi Diaconis, Susan Holmes, and Richard Montgomery, Dynamical bias in the coin toss, *SIAM Review* 49 (2007) 211–223.

For a non-technical summary see Persi Diaconis, Susan Holmes, and Richard Montgomery, The fifty-one percent solution, *What's Happening in the Mathematical Sciences* 7 (2009) 33–45.

Similar effects occur for dice—in fact, not just for the usual cube but for any regular polyhedron. See J. Strzalko, J. Grabski, A. Stefanski, and T. Kapitaniak, Can the dice be fair by dynamics?

International Journal of Bifurcation and Chaos 20 No. 4 (April 2010) 1175–1184.

Eliminating the Impossible $\mathcal{P}$

"Your omission," said Soames, "was to fail to observe that the wine may move, as well as the glasses. I merely pick up the second and fourth glasses, and pour their contents into the seventh and ninth."

Mussel Power

Monique de Jager, Franz J. Weissing, Peter M. J. Herman, Bart A. Nolet, and Johan van de Koppel. Lévy walks evolve through interaction between movement and environmental complexity, *Science* 332 (4 June 2011) 1551–1553.

Proof That the World is Round

On page 74 we saw that when calculating average speeds over a fixed distance we should use the harmonic mean, not the arithmetic mean. The harmonic mean also turns up in the estimation of the distance between two airports when the wind speed is taken into account, for a similar but slightly different reason. To get a simple model, assume that the aircraft's speed relative to the air is c, its path is a straight line and the wind blows along that line in a fixed direction with speed w. Assume both c and w are constant. Then $a = c-w$ and $b = c+w$, and we want to estimate d from the *times* r and s. To get rid of w, we first solve for a and b, getting $a = d/r$ and $b = d/s$. Therefore

$$c-w = d/r \qquad c+w = d/s$$

Adding, $2c = d(1/r+1/s)$. So $c = d(1/r+1/s)/2$. If there had been no wind, a single trip would have taken time t, where $d = ct$. Therefore

$$t = d/c = d/[d(1/r+1/s)/2] = 1/[(1/r+1/s)/2]$$

which is the harmonic mean of r and s.

In short: if we are working in units of aircraft-hours, then this simple model of the effect of wind implies that we should use the harmonic mean of the travel times in the two directions.

123456789 Times X Continued

$$123456789 \times 10 = 1234567890$$
$$123456789 \times 11 = 1358024679$$
$$123456789 \times 12 = 1481481468$$
$$123456789 \times 13 = 1604938257$$
$$123456789 \times 14 = 1728395046$$
$$123456789 \times 15 = 1851851835$$
$$123456789 \times 16 = 1975308624$$
$$123456789 \times 17 = 2098765413$$
$$123456789 \times 18 = 2222222202$$
$$123456789 \times 19 = 2345678991$$

These multiples have all *ten* digits 0–9 in some order, *except* when we multiply by something divisible by 3 ... Until we get to 19, when the pattern stops (19 is not a multiple of 3: answer has two 9's and no 0).

But it picks up again:

$$123456789 \times 20 = 2469135780$$
$$123456789 \times 21 = 2592592569 \qquad \text{(21 is a multiple of 3,}$$
$$\text{so repetitions OK)}$$
$$123456789 \times 22 = 2716049358$$
$$123456789 \times 23 = 2839506147$$

The next exceptions occur at 28 and 29. It works for 30–36, and then fails for 37. I stopped calculating at that point. What's going on here? I have no idea.

The Riddle of the Golden Rhombus

Soames finished tightening the knot, flattened it, and held it up to the light.

"Why, it is a pentagon!" I cried.

"More precisely, Watsup, it *appears* to be a regular pentagon with one diagonal visible and three more hidden. Observe the

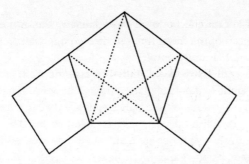

The flattened knot (dotted lines show hidden edges)

absence of a diagonal line running horizontally. Were that to be added, for instance by folding the strip one more time, we would observe—"

"A five-pointed star! A pentacle! Used in black magic to conjure demons!"

Soames nodded. "But without that final fold, and therefore having one edge missing, the pentacle is incomplete and the demon will escape. So the symbol represents a threat to unleash demonic forces upon the world." He gave a humourless smile. "Of course, there are no demons in a supernatural sense, and they can neither be conjured nor unleashed. But there are humans of a demonic disposition—"

"Such as the Al-Jebra terrorist organisation!" I cried. "They have pursued me from Al-Jebraistan with weapons of maths instruction!"

"Calm yourself, Watsup. No, the organisation I have in mind is the Mathemagical Association of Numerica. It is an obscure group, which I strongly suspect to be a front for one of Mogiarty's devilish schemes. I have encountered it before, and I now have the final link in the chain to strike a blow against the foul Professor and destroy this part of his worldwide web of criminality forever. Provided..."

"Provided what, Soames?"

"Provided that we can offer incontrovertible proof when the

case comes to Court. How do we *know* that the pentagon is regular?"

"Is that not absurdly simple?"

"On the contrary, you will shortly be assuring me that it is incredibly subtle and possibly false—though in point of fact the true answer is what one would naively guess. I dare to surmise that once we establish that fact, all else will follow, but the knot's appearance to the eye is not enough. I shall, however, assume that the arrangement of lines in the figure is correct, so we have a pentagon with four of its diagonals. Is it truly regular? That remains to be seen. If true, it must be a consequence of the constant width of the strip.

"Let us, then, label the corners in the manner of the great Euclid of Alexandria, and pursue our geometrical deductions."

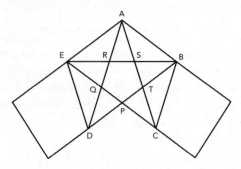

The flattened knot, labelled. Line CD is omitted because we do not yet know whether it is parallel to BE.

I must warn my reader now that the remainder of the discussion will appeal only to those with some facility in Euclidean geometry.

"I begin," said Soames, "with a few simple observations. They can be proved without great difficulty using basic geometry, so I omit the details.

"First, notice that if two strips with parallel edges, having the same width, overlap, then their intersection is a rhombus—a parallelogram with all four sides equal. Moreover, if two such rhombuses have the same width and the same side, they are congruent—they have exactly the same size and shape. The diagram of the flattened knot therefore includes three mutually congruent rhombuses."

"Why only three?" I asked, puzzled.

"Because CD and BE do not coincide with edges of the strip, so we cannot yet say the same of CDRB or DESC. This is why I have not drawn line CD."

I had not noticed that. "It is incredibly subtle, then, Soames. In fact, it might even be false!"

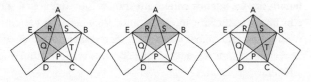

Three congruent rhombuses in the flattened knot

He sighed, I know not why. "Now we come to the central point in my deductions. The diagonals of a rhombus bisect its angles, and opposite angles are equal." Soames marked four of the angles with the Greek letter θ (theta), see the left-hand figure.

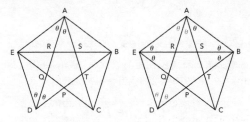

Left: Four equal angles. *Right*: Five more angles, all equal to the first four (in grey).

"For similar reasons, angle CAB is also equal to θ. Since rhombuses DEAT and PEAB are congruent, I can mark four more angles with θ. This leads to the right-hand figure.

"Now, Watsup: what instantly springs to mind?"

"There are an awful lot of θ's," I replied at once.

He grimaced and I heard a low growl in his throat, I know not why. "It is as plain as the neck on a very tall giraffe, Watsup! Consider triangle EAB."

I considered it, initially without enlightenment. Well...the triangle *also* had a lot of θs. In fact...*all* of its angles were composed of θ's! Now I saw it. "The angles of a triangle add to 180°, Soames. In this triangle, the angles are θ, θ, and 3θ. So their sum 5θ is equal to 180°, whence θ = 36°."

"We will make a geometer of you yet," said he. "Now the rest of the proof is easy. The lines DE, EA, AB, and BC are of equal length, being sides of congruent rhombuses. The angles ∠DEA, ∠EAB, and ∠ABC are all equal, since they occur in congruent rhombuses, and one of them is ∠EAB, equal to 3θ, which is 108°. Therefore *all three angles* are 108°. But this is the interior angle of a regular pentagon."

"Therefore DEABC are the corners of a regular pentagon, and I can complete the figure by drawing side CD!" I cried. "How absur—" I caught his eye. "Er, elegant, Soames!"

He shrugged. "A trifle, Watsup. Enough to wreck the Mathemagical Association of Numerica and cause Mogiarty some momentary annoyance. The man himself...I fear he will prove a far harder nut to crack."

Why do Guinness Bubbles go Downwards?

E.S. Benilov, C.P. Cummins, and W.T. Lee. Why do bubbles in Guinness sink? arXiv:1205.5233 [physics.flu-dyn].

The Dogs That Fight in the Park

"The dogs took 10 seconds to collide," Soames declared.

"I shall take your word for it," said I. "But, merely to satisfy my curiosity, how did you arrive at that figure?"

"The problem is symmetric, Watsup, and symmetry often simplifies reasoning. The three dogs are always at the corners of an equilateral triangle. This rotates and shrinks, but keeps its shape. Therefore, from the viewpoint of one of the dogs—say A—it is always running straight towards the next dog B."

"Does not the triangle *rotate*, Soames?"

"Indeed it does, but that is irrelevant, for we may perform the calculation in a rotating frame of reference. What matters is how fast the triangle *shrinks*. Dog B is always running at 60° to the line AB, because the dogs always form an equilateral triangle. So the component of its speed in the direction of dog A is $1/2 \times 4 = 2$ yards per second. Therefore A and B are approaching each other at a combined speed of $4 + 2 = 6$ yards per second, and cover the initial separation of 60 yards in $60/6 = 10$ seconds."

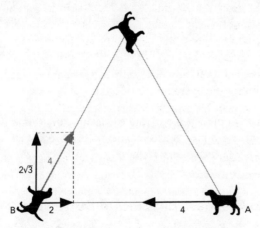

What dog B does in the frame of reference of dog A

Why Do My Friends Have More Friends Than I Do?

Suppose the social network has n people, and person i has x_i friends. Then the average number of friends, over all members, is

$$a = \frac{x_1 + \cdots + x_n}{n}$$

To think about column 3 in the table, the weighted average of how many friends each friend j of person i has, we use a standard mathematical trick and work on person j instead. They turn up as a friend to x_j people—namely, their own friends—and they contribute x_j to the total for each of those friends. So the cases for which person j occurs as a friend contribute x_j^2 to the total. The number of entries in column 3 is $x_1 + \ldots + x_n$. So the weighted average of how many friends each friend has is

$$b = \frac{x_1^2 + \cdots + x_n^2}{x_1 + \cdots + x_n}$$

I claim that for any choices of the x_j we always have $b > a$, unless all the x_j are equal, in which case $b = a$. This follows from a standard inequality relating the average to what engineers call the 'root mean square' (square root of the average of the squares):

$$\frac{x_1 + \cdots + x_n}{n} \leq \sqrt{\frac{x_1^2 + \cdots + x_n^2}{n}}$$

with equality only when all x_j are equal. Squaring this and rearranging, we get $a < b$ except when all x_j are equal, as required. For further information, see:

> https://www.artofproblemsolving.com/Wiki/index.php/
> Root-Mean_Square-Arithmetic_Mean-Geometric_Mean-
> Harmonic_mean_Inequality

The Adventure of the Six Guests

Soames's remark is an example of Ramsey Theory, a branch of combinatorics named after Frank Ramsey, who proved a more general theorem of a similar kind in 1930. His brother Michael became Archbishop of Canterbury. Let's work up to it gently. Suppose that a number of people are seated round a table, with everyone being connected to everyone else by either a fork or a knife. Pick any two numbers f and k. Then there is some number R, depending on f and k, such that if at least R people are present then either f of them are joined by forks, or k by knives.

The smallest such R is denoted by $R(f,k)$ and called the Ramsey number. Soames's proof shows that $R(3,3) = 6$. Ramsey numbers are extraordinarily difficult to calculate, except in a few simple cases. For example, it is known that $R(5,5)$ lies between 43 and 49, but the exact value remains a mystery.

Ramsey proved a more general theorem in which the number of types of connection (knife, fork, whatever—colours are a more common image, but Soames works with whatever is to hand) can be any finite number. The only known non-trivial Ramsey number for more than two types of connection is $R(3,3,3)$, which is 17.

There are innumerable generalisations of this idea. In very few cases is the exact number concerned known. The paper that started it all off is: F.P. Ramsey, On a problem of formal logic, *Proceedings of the London Mathematical Society* 30 (1930) 264–286. As the title suggests, he was thinking about logic, not combinatorics.

Graham's Number

R.L. Graham and B.L. Rothschild, Ramsey theory, *Studies in Combinatorics* (ed. G.-C. Rota) Mathematical Association of America 17 (1978) 80–99.

The Affair of the Above-Average Driver

In 1981 O. Svenson surveyed 161 Swedish and American students, asking them to rate their driving ability and safety relative to the other subjects. For ability, 69% of Swedes considered themselves to be above the median; for safety, the figure was 77%. The figures for the American students were 93% for ability and 88% for safety. Having passed two American driving tests, one of which did not involve getting in the car, I can see why they overestimated their abilities to such an extent. See O. Svenson, Are we all less risky and more skillful than our fellow drivers? *Acta Psychologica* 47 (1981) 143–148.

This effect occurs for many other traits—popularity, health, memory, job performance, even happiness in relationships. It's not especially surprising: it's one way people maintain their own self-esteem. And poor self-esteem can be a sign of psychological inadequacy—so in order to be happy and healthy we have evolved to overestimate how happy and healthy we are.

Dunno about you, but I'm feeling *great*.

The Baffleham Burglary 🔍

"The numbers are 4 and 13," said Soames.

"How utterly amazing. I—"

"You know my methods, Watsup."

"Nevertheless, I find it totally remarkable that you can deduce the answer from such a vague conversation."

"Hmm. We shall see. The essence, Watsup, is that each statement we make adds extra information that we *both* know. And *know* we both know, and so on. Suppose the product of the two numbers is p and their sum is s. Initially you know p, whereas I know s. We each know that the other knows what he knows, but we do not know what that is.

"Since you do not know the two numbers, p cannot be a product of two primes, such as 35. For this is 5×7, and there is no other way to express it as a product of numbers greater than 1, so you would immediately deduce the two numbers. For similar reasons it cannot be the cube of a prime, such as $5^3 = 125$, since this factorises only as 5×25."

"Yes, I see that," I replied.

"More subtly, p cannot be equal to qm where q is prime and m is composite, provided that whenever d divides m and is greater than 1, qd is greater than 100."

"Yeeeesss . . ."

"For example, p cannot be $67 \times 3 \times 5$, which factorises in three ways: 67×15, 201×5, and 335×3. Since the last two use numbers greater than 100, they can be ignored, leaving only 67 and 15 as the two numbers."

"Ah. Quite."

"Now, your remark makes me aware of those facts, but at that point I have already deduced the same information from my sum. In fact, s is not the sum of two such numbers. But you then become aware of that fact, because I tell you, so you then know what to you is new information about s. Of course, we both have to bear in mind that if $s = 200$ then the numbers must both be 100, and if $s = 199$ they are 100 and 99."

"Of course."

"Once you have eliminated the impossible . . . " said Soames, "all that remains turns out to be sums s equal to one of the numbers 11, 17, 23, 27, 29, 35, 37, 41, 47, 51, or 53."

"But previously you have made scathing remarks about—"

"Oh, it works well enough in *mathematics*," he said airily. "For there we can be confident that the impossible really *is* impossible.

"Now, at the relevant stage in the deduction, we *both* know what I have just told you. At which juncture you promptly announce that you can deduce the numbers! So I quickly run through all possible pairs of numbers with those sums, and I find that ten of the eleven possibilities for s share a possible product with a *different* value of s. Since you have told me you now know the numbers, all ten can be eliminated from our investigation. Which leaves only one possible sum, 17, and only one product not occurring for two different values of s. Namely, 52, which arises when we split 17 as $4 + 13$, and only in that manner. Therefore the two numbers must be 4 and 13."

I congratulated him on his perspicacity.

"Send a Baker Street Irreducible to Roulade with this message," he instructed, scribbling the numbers on a scrap of paper. "He will have two arrests within the hour."

Malfatti's Mistake

In 1930 Hyman Lob and Herbert Richmond proved that the greedy algorithm is better than Malfatti's arrangement in some cases. Howard Eves noticed in 1946 that for an isosceles triangle with a very sharp apex, the stacked solution has almost twice the area of the Malfatti arrangement. In 1967 M. Goldberg proved that the greedy algorithm always does better than Malfatti's arrangement. In 1994 Victor Zalgaller and G.A. Los proved that it always produces the largest area.

How to Stop Unwanted Echoes

M.R. Schroeder, Diffuse sound reflection by maximum-length sequence. *Journal of the Acoustical Society of America* 57 (1975) 149–150.

The Enigma of the Versatile Tile

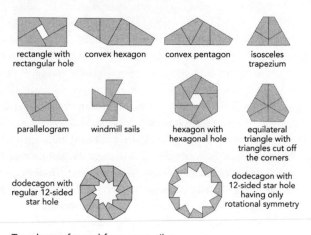

rectangle with rectangular hole

convex hexagon

convex pentagon

isosceles trapezium

parallelogram

windmill sails

hexagon with hexagonal hole

equilateral triangle with triangles cut off the corners

dodecagon with regular 12-sided star hole

dodecagon with 12-sided star hole having only rotational symmetry

Ten shapes formed from versa-tiles

The Thrackle Conjecture

János Pach and Ethan Sterling, Conway's conjecture for monotone thrackles, *American Mathematical Monthly* 118 (June/July 2011) 544–548.

A Tiling That Is Not Periodic

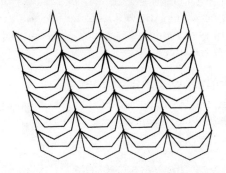

How to tile
periodically
using the 7-gon

The Two Colour Theorem 🔍

Having racked my brains for three hours, I begged Soames to
reveal his secret.

"But then you will tell me how absurdly simple it is."

"Nay! Never!"

"I beg to differ, Watsup. Because for once it *is* absurdly
simple." Silence stretched until he relented. "Very well. Assume
that the only available colours are black and grey, with white for
regions as yet unconsidered. Let us begin by colouring one
region black (see top left figure on page 294). Then I choose one
adjacent region, and colour it grey (top middle). Then I colour an
adjacent region black, then another grey, and so on."

"I see that after the first choice, successive choices are
forced," I said hesitantly.

"Yes! The solution, if it exists, must be *unique*—save for
interchanging the two colours. And you see that eventually the
whole map is coloured, using only black and grey. So in this case,
at least, a solution does exist."

"Agreed. But I don't entirely see—"

"Why. An excellent remark. For once, my dear Watsup, you
have hit the nail firmly on its head, not your own thumb. The
problem is to prove that any such chain of colourings leads to
the same outcome, yes? Because that way, the process can never

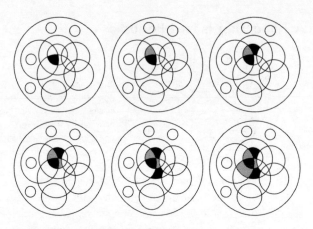

The first few stages in colouring the map

terminate with a situation for which the next remaining region cannot be coloured."

"I think I see that."

"It can be done," said Soames. "But there is a simpler method. Observe that every time we cross a common boundary, the colour changes. Therefore if we cross an odd number of boundaries, we must choose grey, and if we cross an even number of boundaries, we must choose black."

I nodded. "But...how can we be sure that there is no inconsistency?" I blurted.

Soames gave a brief grin. "Because we can take a hint from what I have just said, and prescribe the exact colour of every region. Merely count how many circles contain a given point— not on any circumference, of course, because we do not colour those. If that number is even, colour the point black; if odd, colour it grey.

"Now, crossing any boundary line either adds one containing circle, or subtracts one. Either way, odd changes to even and even to odd, so the two colours on either side of that boundary are different."

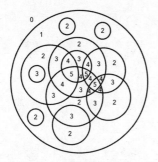

Numbering regions according to how many circles they lie inside. Note how parity (odd/even) changes across boundaries.

The proof was as clear as daylight. "Why, Soames—"

"Of course," he interrupted, with the barest hint of a smile, "some of the circles may be tangent to others. But the same method still applies, suitably interpreted. One must avoid crossing a boundary line at a point of tangency, and a little thought shows that this can always be done."

Well, maybe not quite daylight, but...yes, I understood. "It is—" I began; then paused, seeing his expression. "Very clever," I finished.

The Four Colour Theorem in Space

Four equal spheres can be arranged so that each touches the other three. Place three to make an equilateral triangle, touching each other, and then put the fourth on top to fit into the central dimple, making a tetrahedron. A smaller sphere, of exactly the right size, can now be placed in the middle to touch all four. So we have five spheres, each touching the other four, and these must all have different colours.

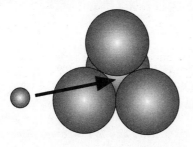

Fitting in the fifth sphere

The Greek Integrator

Answer first. We have to solve the equation $\frac{4}{3}\pi r^3 = 4\pi r^2$. Divide by $4\pi r^2$ to get $\frac{1}{3}r = 1$. Therefore $r = 3$.

Now for the palimpsest.

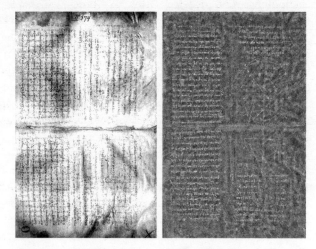

Left: A typical page of the Archimedes palimpsest. The thirteenth-century religious text runs vertically; the fainter original runs horizontally. *Right*: Cleaned-up page with clear mathematical diagrams.

Archimedes' original manuscript has not survived, but this copy (no doubt the culmination of a series of copies) was made by a Byzantine monk around AD 950. In 1229 it was unbound and scrubbed (fairly) clean, along with at least six other manuscripts. The sheets were folded in half and used to write a 177-page Christian liturgical text—a description of the procedures for religious services.

In the 1840s Constantin von Tischendorf, a German Biblical scholar, came across this text in Constantinople (now Istanbul), noticed faint mathematical writings, and brought a page of it home. In 1906 a Danish scholar, Johan Heiberg, realised that part of the palimpsest was a work of Archimedes. He photographed it, publishing some extracts in 1910 and 1915. Thomas Heath translated the material shortly afterwards, but it attracted little attention. In the 1920s the document was in the possession of a French collector; by 1998 it had somehow made its way to the USA, becoming the subject of a court case between Christie's auction house and the Greek Orthodox Church, which alleged that the document had been stolen from a monastery in 1920. The judge ruled in Christie's favour on the grounds that the delay between the alleged theft and the legal action had been too long. The document was purchased by an anonymous buyer (reported by *Der Spiegel* to be Amazon's founder Jeff Bezos) for \$2 million. Between 1999 and 2008 the document was conserved at the Walters Art Museum, Baltimore, and analysed by a team of imaging scientists to enhance the hidden writing.

Archimedes' method can be explained (using modern language and symbolism) as follows. Start with a sphere of radius 1, its circumscribed cylinder, and a cone. If we place the centre of the sphere at position $x = 1$ on the real line, then the cross-sectional radius at any x between 0 and 2 is $\sqrt{x(2-x)}$, and its mass is proportional to π times the square of this, namely $\pi x(2-x)$ $= 2\pi x - \pi x^2$.

Next, consider a cone obtained by rotating the line $y = x$ around the x-axis, again for $0 \le x \le 2$. The cross section at x is a circle of radius x, and has area πx^2. Its mass is proportional to this, with the same constant of proportionality, so the combined mass of the slice of sphere and the slice of the cone is $(2\pi x - \pi x^2) + \pi x^2 = 2\pi x$.

Place the two slices at $x = -1$, distance 1 to the left of the origin. By the law of the lever, they are exactly balanced by a circle of radius 1 placed distance x to the right.

Now move all slices of the sphere and the cone to the *same* point $x = -1$, so that their total mass is concentrated at this single point. The corresponding (and balancing) circles all have radius 1, and are placed at all distances from 0 to 2. They therefore form a cylinder. Its centre of mass is in the middle, at $x = 1$. Therefore, by the law of the lever,

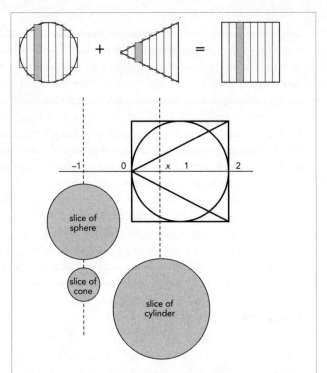

What Archimedes did. *Top*: Slice a sphere, cone, and cylinder (shown in cross section: sphere = circle, cone = triangle, cylinder = square) like loaves of sliced bread. Then the volume of a slice of the cylinder (grey) is the sum of the volumes of the corresponding slices of the sphere and the cone. The slices here have nonzero thickness, which introduces errors. Archimedes thought about infinitely thin slices, for which the errors become as small as we please. *Bottom*: the weighing argument that relates the three volumes. Slices at x of sphere and cone, placed at -1, balance slice of cylinder placed at x.

mass of sphere + mass of cone = mass of cylinder

and since mass is proportional to volume,

volume of sphere + volume of cone = volume of cylinder.

However, Archimedes already knew that the volume of the cone is one third that of the cylinder (one third area of base times height, remember?), so the volume of the sphere is two thirds that of the cylinder. The volume of the cylinder is the area of the base (πr^2) times the height ($2r$), that is, $2\pi r^3$. So the volume of the sphere is $\frac{2}{3}$ of this, namely $\frac{4}{3}\pi r^3$.

Archimedes derived the surface area of the sphere by a similar procedure.

He described the process geometrically, but it's easier to follow the argument using modern notation. Considering that he did all this around 250 BC, and that he also developed the law of the lever, it's an amazing achievement.

Why the Leopard Got Its Spots

W.L. Allen, I.C. Cuthill, N.E. Scott-Samuel, and R.J. Baddeley. Why the leopard got its spots: relating pattern development to ecology in felids, *Proceedings of the Royal Society B: Biological Sciences* 278 (2011) 1373–1380.

Polygons Forever

It looks as though the figure will grow without limit, but actually it remains within a bounded region of the plane: a circle whose radius is about 8.7.

The ratio of the radii of a circle circumscribed about a regular *n*-gon and one inscribed in it is $\sec \pi/n$, where sec is the trigonometric secant function and I'm using radian measure for the angle. (Replace π by $180°$ for degree measure.) So for each *n* the radius of the circle circumscribed about the regular *n*-gon in the picture is

$$S = \sec \pi/3 \times \sec \pi/4 \times \sec \pi/5 \times \cdots \times \sec \pi/n$$

We want the limit of this product as *n* tends to infinity. Take logarithms:

$$\log S = \log \sec \pi/3 + \log \sec \pi/4 + \log \sec \pi/5 + \cdots + \log \sec \pi/n$$

When *x* is small, $\log \sec x \sim x^2/2$, so this series can be compared with

$$1/3^2 + 1/4^2 + 1/5^2 + \cdots + 1/n^2$$

which converges as *n* tends to infinity. Therefore log *S* is finite, so *S* is finite. The sum of the terms up to *n* = 1,000,000 yields 8.7 as a reasonable estimate.

I learned about this problem, and the answer given above, from a book review by Harold Boas [*American Mathematical Monthly* 121 (2014) 178–182]. He traced it back as far as *Mathematics and the Imagination* by Edward Kasner and James Newman in 1940. He writes: "Perhaps this amusing example will become part of the standard lore if the figure gets reproduced in enough books."

I'm trying, Harold.

The Adventure of the Rowing Men

Soames and I found two further arrangements, not counting reflections in the centreline as different:

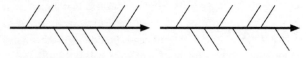

Arrangements 0167 and 0356

"For all the mechanical complexity of the problem," said Soames, "it reduces ultimately to mere arithmetic. We have to split the numbers from 0 to 7 into two sets, each with sum 14."

"If we know one such set, the other is determined and also has sum 14."

"Yes, Watsup, that is evident: list the numbers that are not in the first set."

"I agree it is trivial, Soames, but it implies that we can work on the set containing 0, which amounts to putting the stern oar on the left, which we may assume by applying a reflection if necessary. Thereby reducing the number of cases to be considered."

"True."

Now the deductions almost made themselves. "If the set also contains 1," I pointed out, "then the other two add to 13, so they *must* be 6 and 7, giving 0167. If it does not contain 1 but does contain 2, then the only possibility is 0257. If it starts 03 there are two possibilities: 0347 and 0356. We can dismiss arrangements that start 04 because it is not possible to make 10 from two of the numbers 5, 6, 7. A similar argument disposes of 05, 06, and 07."

"So you have deduced," Soames summarised, "that the only possibilities, excluding left–right reflections, are

    0167   0257   0356   0347

Now 0257 is the German rig and 0347 the Italian. There are two others: those that I have laid out with my matchst—"

He suddenly sat up, rigid. "Great Scott!"

"What, Soames?"

"It has just struck me, no pun intended, Watsup, that this match—" he waved it at me— "is not a rare early Congreve, as I had imagined, but one of Irinyi's noiseless matches. When his chemistry professor blew himself up, Irinyi was inspired to replace potassium chlorate by lead dioxide in the head of the match."

"Ah. Is that significant, Soames?"

"Most assuredly, Watsup. It casts an entirely new light— again no pun intended—on one of our most bizarre unsolved cases."

"The Remarkable Affair of the Upside Down Teapot!" I cried.

"You have it, Watsup. Now, if your notes record whether the match was dropped to the left or the right of the mummified parrot . . . "

Soames's analysis is based on:

Maurice Brearley, 'Oar arrangements in rowing eights', in *Optimal Strategies in Sports* (ed. S.P. Ladany and R.E. Machol), North-Holland 1977.

John Barrow, *One Hundred Essential Things You Didn't Know You Didn't Know*, W.W. Norton, New York 2009.

As Soames warned, it is an initial simplified approach to a highly complex issue.

By the way, the 1877 Boat Race was a dead heat—the only one in the history of the event.

Rings of Regular Solids

John Mason and Theodorus Dekker found simpler methods than Świerczkowski's to prove the impossibility. Whenever you glue two identical tetrahedrons by their faces, each is a reflection of the other in the common face.

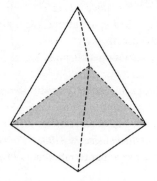

Two tetrahedrons with a common face (shaded). Each is a reflection of the other in this face.

Start with one tetrahedron. It has four faces, so there are four such reflections; call them r_1, r_2, r_3, and r_4. Each reflection sends everything back where it started if you do it twice, so $r_1 r_1 = e$, where e is the transformation 'do nothing'. The same goes for the other reflections. So all combinations of several reflections are products like

$$r_1 r_4 r_3 r_4 r_2 r_1 r_3 r_1$$

where the sequence of subscripts 14342131 can be any sequence formed by the four numbers 1, 2, 3, 4 in which no number ever occurs twice in a row. For example, 14332131 is not permitted. The reason is that here $r_3 r_3$ is the same reflection performed twice, so it is the same as e, which has no effect and can therefore be deleted.

If a chain closes up, one further reflection applied to the last tetrahedron in the chain produces a tetrahedron that coincides with the initial one. So we get an equation like

$$r_1 r_4 r_3 r_4 r_2 r_1 r_3 r_1 = e$$

(only longer and more complicated) where e stands for 'do nothing'. By writing down formulas for the four reflections, and using suitable algebraic methods, it can be proved that no such equation holds. For details, see:

T. J. Dekker, On reflections in Euclidean spaces generating free products, *Nieuw Archief voor Wiskunde* 7 (1959) 57–60.

M. Elgersma and S. Wagon, Closing a Platonic gap, *Mathematical Intelligencer* in the press.

J. H. Mason, Can regular tetrahedrons be glued together face to face to form a ring? *Mathematical Gazette* 56 (1972) 194–197.

H. Steinhaus, Problem 175, *Colloquium Mathematicum* 4 (1957) 243.

S. Świerczkowski, On a free group of rotations of the Euclidean space, *Indagationes Mathematicae* 20 (1958) 376–378.

S. Świerczkowski, On chains of regular tetrahedra, *Colloquium Mathematicum* 7 (1959) 9–10.

The Impossible Route

"As you so rightly say, you do not see it," said Soames. "You know my methods: use them."

"Very well, Soames," I replied. "You have always instructed me to discard that which is irrelevant. I shall therefore repeat my analysis, and to eliminate any conceivable possibility of error I shall represent the problem in its simplest form. I number the regions on the map—like so. There are five of them. Then I draw a diagram—I believe it is called a *graph*—showing the regions and their connections in schematic form."

He remained silent, his expression unreadable.

"We must proceed from region 1 to region 5, leaving bridge A to the last. Starting from 1, the only alternative is to cross bridge B, and then C and D are forced upon us. We must use either bridge E or F. Let us say we take E. We cannot use F because that

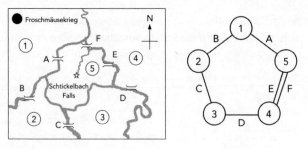

Left: Watsup's five regions. *Right*: Graph of connections.

takes us to region 4 and we cannot proceed further. However, we cannot then use A, because that takes us to region 1and we cannot proceed further. The same goes if we employ F in place of E. I rest my case."

"Why, Watsup?"

"Because, Soames, I have eliminated the impossible." He raised one eyebrow. "So what remains, however unlikely," I continued, "must be—"

"Go on."

"But Soames, *nothing* remains! Therefore the problem has no solution!"

"Wrong. I have told you that there are eight."

"Then you must have lied about the conditions."

"I did not."

"Then I am stumped. What have I missed out?"

"Nothing."

"But—"

"You put too much *in*, Watsup. You made an unwarranted assumption. Your error was to assume that the path does not leave the map."

"But you told me that the rivers continue to flow to the Swiss borders, and we are not permitted to recross the border."

"Yes. But the map does not depict the whole of Switzerland. Where does the river come from?"

"*D'oh!*" I struck my forehead with my hand.

"Dough?"

"Merely an inadvertent expression berating myself for my own stupidity, Soames. Not 'dough'. More along the lines of 'D'oh!' "

"I advise you to avoid it, Watsup. It does not become you, and it will never catch on."

"As you say, Soames. What caused my outburst was the realisation that we can complete my second attempt by encircling the source of the river and passing over bridge A."

"Correct."

"So regions 1 and 4 in my figure are actually the same region."

"Indeed."

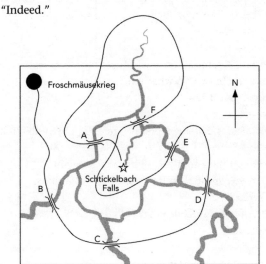

Soames's route

"That," I said after a moment, "was unfair. How am I to know that the river rises within Swiss borders? The source was not shown on your map."

"Because, Watsup, I told you that there is at least one route satisfying my conditions. It follows that the source *must* lie in Switzerland."

Touché. Then I remembered that he had referred to eight routes. "I see a second route, Soames: interchange bridges E and F. But I confess the other six elude me."

"Ah. Your assertion that we must begin with bridge B is no longer valid when regions 1 and 4 are merged. Let me redraw your simplified figure correctly."

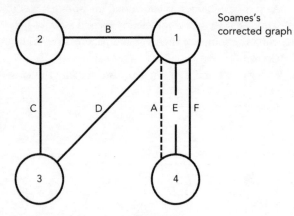

Soames's corrected graph

"I have drawn bridge A as a dotted line as a reminder that we must leave A until last. Observe: starting from region 1, the bridges other than A form two distinct loops: BCD and EF. We can traverse each loop in two directions: BCD or DCB, and EF or FE. Furthermore, we can start with either loop and then traverse the other. Finally, we must append bridge A. So the different routes are

BCD–EF–A	DCB–EF–A	BCD–FE–A	DCB–FE–A
EF–BCD–A	EF–DCB–A	FE–BCD–A	FE–DCB–A

"A total of eight."

"I see my error clearly now, Soames," I admitted.

"You see your *specific* error, Watsup, but not the underlying generality, which afflicts all arguments about eliminating the impossible. "

I shook my head, puzzled. "What do you mean?"

"I mean, Watsup, that you did not consider *all* possibilities. And the reason was—"

Again I struck my head with my hand, but this time I refrained from uttering a sound, not wishing to be the butt of Soames's scorn. "I forgot to think outside the box."